TO LIVE TO WORK

JANICE C. H. KIM

# To Live to Work

*Factory Women in Colonial Korea, 1910–1945*

STANFORD UNIVERSITY PRESS

STANFORD, CALIFORNIA

Stanford University Press
Stanford, California
©2009 by the Board of Trustees of the Leland
Stanford Junior University. All rights reserved.

Printed in the United States of America on acid-free,
archival-quality paper

Library of Congress Cataloging-in-Publication Data

Kim, Janice C. H.
    To live to work : factory women in colonial Korea,
1910–1945 / Janice C. H. Kim.
        p. cm.
    Includes bibliographical references and index.
    ISBN 978-0-8047-5909-0 (cloth : alk. paper)
        1. Women—Employment—Korea—History—
20th century.   2. Factory system—Korea—
History—20th century.   I. Title.
    HD6068.2.K6K54 2009
    331.4'80951909041—dc22              2008038351

Typeset by Westchester Book Group in 10/14 Janson

*For Cho Jai Seong*

# Contents

*Illustrations*                                                                 ix

*Figures and Tables*                                                            xi

*Preface*                                                                       xiii

Introduction                                                                    1

1.    Locating Korean Factory Women in Time and Place                 26

2.    Modernization and the Rise of Women's Wage Work                 50

3.    Lives and Labors Inside the Factories                           75

4.    Contests of Power and Workers' Modes of Association             101

5.    The Pacific War and the Life Courses of Working Women           127

      Conclusions: The Legacies of Colonial Working Women   155

*Notes*                                                                         177

*Bibliography*                                                                  221

*Index*                                                                         243

# Illustrations

Plate 1.1   An open-air cotton storage in Mokp'o (1920s)     41

Plate 1.2   Cotton market in Seoul (1930s)     41

Plate 1.3   Downtown Seoul (1920s)     42

Plate 1.4   Downtown Wŏnsan (1930s)     42

Plate 1.5   The Ch'ungmuro (Honmachi) shopping district in Seoul (1930s)     43

Plate 1.6   Weaving class in a school for girls (1920s)     48

Plate 2.1   Farm women selecting cotton (1930s)     58

Plate 2.2   Silk reeling in a household in colonial Korea     59

Plate 2.3   Cotton spinning in a household in colonial Korea     60

Plate 2.4   Cannery on the east coast (1920s)     66

Plate 3.1   Cotton ginning or "scutching and mixing" workshop (1920s)     83

Plate 3.2   Novices piecing thread in a spinning room (1920s)     84

Plate 4.1   Weaving room (1930s)     109

Plate 4.2   Silk-reeling workers (1930s)                                    112

Plate 5.1   Rural women mobilized for agricultural labor during the
            Pacific War (1940s)                                            138

Plate 5.2   Patriotic women's associations in late-colonial Korea (1940s)   139

Plate 5.3   Fencing practice in a school for girls (1940s)                  139

Plate 5.4   Wartime students write "Do away with rats" (1940s)             142

# Figures and Tables

## Figures

Figure 1.1  Map of Korea (1930)                                    xviii

Figure 4.1  Demands of the Pyongyang Amalgamated Rubber Worker's
            Union (P'yŏngyang komu chikkong chohap) (1930)         107

Figure 4.2  Demands Presented to the Owners of the Chosŏn Spinning and
            Weaving Company (Chōsen boseki kabushiki kaisha) (1930)  108

Figure 4.3  Demands of the Workers of the Pyongyang Taedong Hosiery
            Factory (P'yŏngyang Taedong yangmal kongjang) (1932)   110

Figure 5.1  Map of East Asia (1945)                                147

## Tables

Table 1.1  Interviews                                              22

Table 2.1  Distribution of Male and Female Workers Across
           Industries (1921–43)                                   65

Table 5.1  Chronology of Labor-Related Orders for the War Effort
           (1937–45)                                              134

## Preface

My interest in the Korean past was sparked over a decade ago with a simple exchange during office hours, when I asked my Chinese history professor about possible careers in the discipline. "How is your Korean?" he asked. I replied, "Not bad." He then posed, "How about Korean history?" I exited the meeting with a list of names and titles and headed to the library. Little did I know then that my past in Korea would serve as a stepping-stone to my future. Little did I know then that disentangling the intricacies of Korean history would require a knowledge of the cultures and languages of China and Japan, as well as an understanding of the peninsula's economic and political relations with Russia and the United States. It was also during my undergraduate years at Johns Hopkins that Vernon Lidtke and Jack Greene introduced me to historiography, microhistory, and the "Mary Paul Letters." Ron Walters inspired a love of feminist theory and style, showing me how to extract not only paragraphs but pages, without hesitation. I am grateful for the advice and encouragement provided by many former professors and, in particular, Bill Rowe, which continues to influence my analytical and interpretative outlook.

It was also in Baltimore that I first became intrigued by the correlations between gender, labor, and modernity. Impressed by Joan Scott and Louise Tilly's work among others, I sought to further investigate two discursive themes: "debates among feminists about women and work," and "debates among historians about social and economic change."[1] To grasp the idiosyncrasies of Korean women's history, however, required recognizing the tenacity of Confucianism in modern society and the contested character of

Korea's economic development. External pressures influenced relations among Koreans, and at no point were those forces more prominent than after the institution of Japanese colonialism in 1910. To fully appreciate the "intertwined histories and overlapping territories" that comprised modern Korea, I myself had to cross waters to other territories new and old.[2]

As a graduate student at the School of Oriental and African Studies (SOAS), University of London, I benefited from the input of an impressive array of scholars. I had the good fortune of meeting early in my graduate career Kenneth Wells of Australia National University, who referred me to a range of colonial periodicals, which furnished a background of the workings of gender in colonial Korean society. Research trips to the Library of Congress and the Harvard-Yenching Library during 1997 and 1998 facilitated my acquisition of preliminary sources and offered other leads. Carter Eckert of Harvard University introduced me to mentors in Korea, and it was also at Yenching that I became acquainted with Soon-Won Park's research on colonial labor history. Since, I've considered myself fortunate to have at hand Soon-Won's expertise and guidance. Mentors and colleagues at and around SOAS including Remco Breuker, Frank Dikötter, Keith Howard, Stefan Knoob, Roald Maliangkay, Satona Suzuki, Sem Vermeersch, Carl Young, and the late Ralph Smith were valuable discussants. Nevertheless, this project could not have been accomplished without Martina Deuchler's unyielding interest and assistance. I would like to record my gratitude for her continuous encouragement as well as the standard of quality she inspired.

Much of the research for this book was conducted during 1998 and 1999 when I was a special research student working with An Pyŏngjik at Seoul National University. With his sponsorship, I was given access to the imperial archives of the Seoul National University Library as well as other, now "hard-to-access," collections. Chu Ikchong of Seoul National University and Pak Sŏngch'ŏl of the Kyŏngsŏng Spinning and Weaving Company escorted me to the company archives. These scholars and others affiliated with the Naksŏngdae Institute for Economic History extended much appreciated support in acquiring and comprehending the details of the primary sources.

More extensive source collection would also not have been possible without the assistance of Kim Kyŏngil of the Academy of Korean Studies, Chŏng Chinsŏng of Seoul National University, Chŏng Kŭnsik of Chŏnnam

University, and Yi Chŏngok of Hyosŏng Catholic University. Kang Isu of Sangji University went so far as to lend me two of her interview tapes, which are mentioned in this work. Nonetheless, most of the informants referred to in this study were located with the help of Kim Ŭnsik and Yi Hija of the Society for the Survivors and the Bereaved Families of the Pacific War (T'aep'yŏngyang chŏnjaeng hŭisaengja yujokhoe). Of course, the finer details of working women's histories could not have been recovered without the testimonies of the following former factory workers: Han Kiyŏng, Kang Pokchŏm, Kim Chŏngmin, Kim Chŏngnam, Kim Ŭnnye, Kim Yŏngsŏn, Pak Sun'gŭm, Yi Chaeyun, and Yi Chungnye. I would like to sincerely thank these interviewees, as well as other activists and scholars in Korea, whose generous assistance offered insight and shed light on the everyday experiences of factory women in colonial Korea.

My colleagues in Toronto have been vital sources of stimulation and support during the later stages of revision. Bettina Bradbury made Canada warm, Tom Cohen indulged my questions on syntax and "flow," Josh Fogel furnished practical advice, and Joan Judge, much-needed doses of optimism. Although I am much obliged to these and other historians at York University, I am perhaps most indebted to Bob Wakabayashi, who let me know, more importantly, "what *not* to do." Also invigorating has been having a modern Korean historian in town. Andre Schmid of the University of Toronto, to whom I owe many hours of reading, writing, phone calls, and e-mails in return, has been charitable to a fault with his time, candid feedback, and tireless encouragement. Colleagues who have either directly or indirectly extended feedback include: Charles Armstrong, Donald Baker, Bruce Cumings, Henry Em, Dong-Sook Shin Gills, Merose Hwang, Mihyon Jeon, Ken Kawashima, Seung-Kyung Kim, Dennis McNamara, Janet Poole, Mike Robinson, David Steinberg, Jun Yoo, and the late Jim Palais. Just as significant, acknowledgment is needed for those friends, including Jo Hirst, Jim Soriano, Sophia Darling, Ki Jackson, Jason Blake, Mark Morris, and the late Nupur Srivastava, who listened to and helped me refine my ramblings on silkworms, spindles, and other sedation-inducing subjects.

I am grateful to the sponsors of earlier published articles for allowing me to include portions of them in this book, including "The Pacific War and Working Women in Colonial Korea," *Signs* (Autumn 2007) and "Varieties of Women's Work in Colonial Korea," *Review of Korean Studies* (June 2007).

I am also indebted to the many reviewers who have commented on my work as well as Muriel Bell for her interest in the book's publication. Although I thank colleagues, friends, and family for making this endeavor enjoyable, I alone bear responsibility for its final outcome.

Funding was provided by a number of scholarships and grants awarded by the Association for Asian Studies, the British Council, the Korea Foundation, the Social Sciences and Humanities Research Council of Canada, and the York University Faculty of Arts. The McCune-Reishauer and Hepburn systems for the romanization of Korean and Japanese are used throughout the work.

Although my curiosity about gender and labor in Korea was piqued with the assistance of professors, colleagues, and friends, in retrospect, I see that it originated with the Korean women I knew long before I entered university halls. I am grateful to Cho Jai Kyung, who accompanied me on my interviews and provided me with a home in Korea. Chung Hak Jae taught me to appreciate the subtler nuances of the Korean language. Yang To Dam, the family matriarch, tolerated the inconveniences of my unexpected return to Seoul in the last years of her life. I am most obliged, however, to Cho Jai Seong, who furnished daily examples of the bounds of women's productivity. I thank and, in particular, I owe much to, the precedents of my family in Korea, most notably the women, who have set incredible standards for industriousness and strength. While working for wages, caring for children, cooking meals, and cleaning homes, I remember these women instructing us, the next generation, "Study hard. Work hard. Time is precious. To live is to work!" Indeed, by living to work, my mother afforded me opportunities few could merit, and without her support this project could not have been completed. It is to her, Cho Jai Seong, and to the Korean women of her generation and before, this book is dedicated.

TO LIVE TO WORK

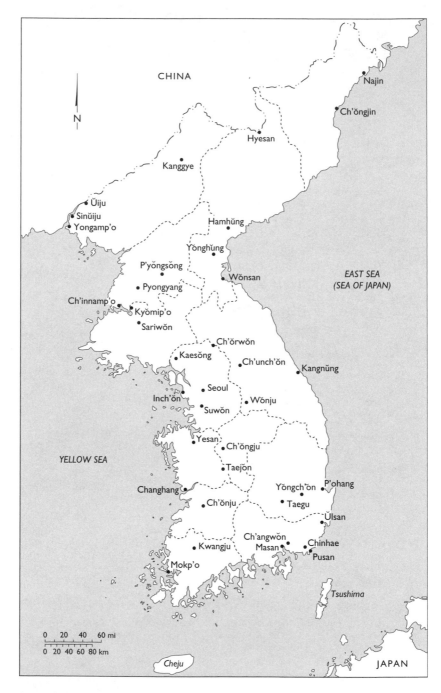

CHINA

N

• Najin

• Ch'ŏngjin

Hyesan

Kanggye

• Ŭiju
• Sinŭiju
• Yongamp'o

Hamhŭng

Yŏnghŭng

P'yŏngsŏng

• Wŏnsan

EAST SEA
(SEA OF JAPAN)

• Pyongyang

Ch'innamp'o
Kyŏmip'o
• Sariwŏn

Ch'ŏrwŏn

Kaesŏng
• Ch'unch'ŏn
Kangnŭng

• Seoul
Inch'ŏn
•Suwŏn
• Wŏnju

Yesan
• Ch'ŏngju

•Taejŏn

YELLOW SEA

Yŏngch'ŏn • P'ohang

Changhang
• Taegu

Ch'ŏnju

Ŭlsan

Ch'angwŏn
Chinhae

• Kwangju
Masan
Pusan

• Mokp'o

Tsushima

0   20   40   60 mi
0   20  40  60  80 km

Cheju

JAPAN

*Figure 1.1*  Map of Korea (1930)

# Introduction

Once contested, it is now widely accepted that Korea's modern economic development began during the Japanese colonial era.[1] Between 1910 and 1945, the population grew from fifteen to twenty-four million. Although agricultural output also increased, the total value of production in mining and manufacturing skyrocketed from approximately thirty to five hundred million yen from 1910 to 1940. Urban migration coincided with population growth and capitalist development; the peninsula's share of city-dwellers jumped from 3 to 14 percent in the thirty-five years of colonial rule. Rural Koreans were drawn to the cities not only for jobs but also for convenient transportation networks, banks, hospitals, schools, department stores, and cinemas among other opportunities, benefits, and attractions. Mechanization and urbanization, however, were not easily achieved. In describing the drawbacks of modernity, contemporary intellectuals turned to prototypes of impoverished "mill girls." A 1936 article in the *Chosŏn chungang ilbo* (Korea Central Daily) depicts the dismal conditions of working women:

Female factory workers, who work in dim-lit workshops, under the fearful watch of overseers, breathing in the scorching 100 degrees air, working with splintering muscles and shattering bones, are mostly girls from fifteen to sixteen, recruited from the countryside. These women work for fifteen to sixteen chŏn a day for six to seven years and after all the years of exertion, receive only forty to fifty chŏn in the end. Though the residence is called a dormitory, over ten women are put into a tiny room and under the watchful eyes of numerous guards. Freedom is totally suppressed. The work hours are long and the meals are insufficient and due to malnutrition and overwork, the health of women workers declines rapidly. Because of the lack of sunlight and air, these women resemble sufferers of serious illnesses and there are many incidents of workers fainting due to exhaustion. But, by virtue of the stringent regulations of the company, workers endure these conditions for failure to keep with the regulations results in being beaten.[2]

The dramatic juxtaposition of youthful femininity and mechanized labor, brought on by industrialization under Japanese rule, was emphasized by intellectuals and activists in colonial Korea (1910–45). Loaded with political, economic, and cultural implications, such descriptions of factory women triggered fears of the social costs of modernization.

The seeming powerlessness of factory women was a theme filled with political potential. Feminists, for instance, underscored the gender-specific abuses of industrial capitalism and deemed the tradition of patriarchy as responsible for the plight of the factory girl. Nationalists pointed to the ethnically biased management procedures of the largely Japanese-owned manufacturing complex in early twentieth-century Korea, contending that the exploitation of young Korean women was rooted in Japanese colonialism. Marxists, on the other hand, viewed the struggle in terms of socioeconomic class distinctions. Although not mutually exclusive, each association highlighted specific characteristics of women workers to reaffirm its political interests but neglected other, often overlapping and contradictory, traits. The politics of representation relied on a double-edged view of subjecthood. While representation sought to extend visibility to individuals as political subjects, the project of defining individual visibility intrinsically simplified or objectified otherwise plural subjects. By representing them, intellectuals affixed meaning onto factory women's lives, unwittingly perpetuating static models of their experiences and identities.

Contemporary and subsequent scholarly interpretations of early twentieth-century working women were premised on, among others, four theoretical traditions concerning the subject or the self: East Asian paternalism, Enlightenment liberalism, nationalism, and Marxism. Korean Confucianism, although differing from European patriarchy, embraced a logic that placed women in secondary public status. According to East Asian tradition, a woman's personhood was meaningful only when affiliated to her father, her husband, or her son. Early modern European writings, transmitted by Western missionaries, Japanese intellectuals, and Korean émigrés, offered turn-of-the-century Koreans, including women, alternative ideologies for conceiving the self. Drawing on rationalism, Korean intellectuals believed that the human mind mirrored nature, that actions were governed by reason, and that self-determination was possible. Protestant conversion, beginning in the late nineteenth century, fostered alternative visions of femininity by extending women's duties in private, moralistic spheres and supporting the ideals of individual rights, liberties, and values. For colonial Koreans, however, the promise of a realizable self seemed inflated when viewed with the lack of its necessary corollary: citizenship in an autonomous nation-state. Although activists espoused liberalism and nationalism more readily, the writings of nineteenth- and twentieth-century socialists, introduced by Korean émigrés and Japanese radicals, also influenced an important segment of the Korean population. Left-of-center Koreans saw socialism and communism as vehicles of empowerment. Even for Marxists, however, plans for international revolution came second to domestic revolution, achievable only through national liberation. Intellectual renderings of factory women were positional and, from the perspectives of feminists, socialists, and nationalists, factory women were triply discriminated by their gender, class, and nationality. The rationales of colonial elites evidenced an analytical distinction between the subjects of their sympathy and themselves, the "subjects" of the larger political discourse. Rather than being viewed as an individual, the "factory woman" became an object of study, a social category, and a means to narrate the disenfranchisement of coloniality, femininity, and material poverty.

Taking the objects of colonial intellectuals' representation and turning them into subjects complicates such fixed impressions of working women.

Because the person who speaks and acts is always a "multiciplicity," affected by various and manifold impulses, no intellectual, party, or union can represent "those who act and struggle." Drawing on subalternity as first articulated by Antonio Gramsci, Gayatri Spivak poses, "Are those who act and struggle mute, as opposed to those who act and speak?"[3] In colonial Korea, although intellectuals generally spoke *for* factory women, the voices of factory girls were not inaudible. A poem entitled "Women Workers," published in the periodical *Kaebyŏk* (Creation), captures some of the lesser-known meanings of factory work.

> They say that spring has come but I have not yet bid farewell to winter
> An author's affections for an old nation asleep
> The first factory bell at dawn
> But I only see a blue-black sky
> When three hundred thousand people barely start to wake
> The dinner table for my dead husband
> I did not get to set, for I ran
> Hearing the breath of an inner evil
> Inside the goods made in a demon's cave
> With bent knees, not once being able to turn my head
> The thought of spending twelve hours as such makes me shudder
> The dismal gaze of the grotesquely grotesque supervisor
> This damned world where I have to heed that monster. . . .
> Oh ancestors! Oh my husband!
> Why did you leave this dreaded world?[4]

The poem furnishes a glimpse of the intricacies of the factory women's problems as perceived through the lens of their personal histories. These more intimate sources question the conformity of individual and social consciousness: Were factory women's identities indeed defined by feminist, working-class, or nationalist consciousness? Or were their identities more complex, contingent, and transient? Were women assigned to factory work because of their gender, class, and nationality or because of historical, local, and familial circumstances? How did the meanings of women's factory work merge with the narratives of their lives, and how did individuals conceptualize themselves as the protagonists of these stories? More detailed accounts of factory women in early twentieth-century Korea challenge assumptions of the ontological certainty of class, gender, or nationality and the epistemological integrity of the consciousness claimed by each association.

This study offers an overview of the evolution of the female industrial workforce in colonial Korea but, just as important, it is a hermeneutic critique and reevaluation of the meanings of factory labor for women workers themselves. Departing from Enlightenment dichotomizations of self and society, twentieth-century scholarship has shown social structures and the people constituting them to share mutual influence. Many humanists and social scientists no longer view the self as an independent entity but as a constantly modifying construction that gains life through performance in fluctuating social environments. Questions concerning the epistemological foundations of the self also prompted reconceptualizations of individual agency. Whether individual or institutional, power is always confined by circumstance. That power is limited, nonetheless, does not mean that the actions of those with fewer choices are more reactive than active because their decisions often lead to change. The aim of this introduction is to render an alternative framework for understanding factory women in the colonial era and, by extension, female wage workers in twentieth-century Korea.

New hermeneutics, however, requires a review of hermeneutics past. Thus, I examine the historiography of working women in light of colonial nationalist, socialist, and feminist perspectives. Although not mutually exclusive, female factory workers have been invariably linked to sociopolitical associations and their causes. Reconsidering how political and personal identities are formed, however, demonstrates that working women's consciousness was neither as bound nor as fixed as their representatives claimed. In everyday life, ideologies and alliances were not simply reinforced or rejected but constantly compromised. Finally, I elaborate on the procedures of my research and the presentation of the book. Throughout this work, I argue that colonial women workers did not conceive of themselves as ideologues who spoke on behalf of all Korean women. Nevertheless, their separate protests for the improvement of working conditions and treatment as well as their struggles for their families and communities illustrate the extent of popular women's awareness and activism in early twentieth-century Korea.

## Nationalism and Feminism

As Jacques Derrida and others suggest, because women's liberation posed fundamental ontological and epistemological questions concerning historical and contemporary society, women's history was not isolated from the political struggles of modernity.[5] Due to the unavoidable alliance between feminism and politics, studies of women's pasts have often determined female historical agency according to contemporary definitions of civil rights and liberties. The clear and exclusive association between nationalism and feminism in Korean historiography adheres to a tradition of patriotic scholarship pervasive in North and South Korea, where recollections of the past work toward the edification of the nation. North Korean historiography follows Kim Il-Sung's ideology of self-reliance (*chuch'e*), promoting remembrances glorifying the corporate family state.[6] In South Korea, this partiality is less pronounced but a general scholarly tendency to attribute modern social reforms to the efforts of early nationalists often blurs the lines between critical and popular understandings of Korea's modern history. In the republics of North and South Korea, both founded in 1948, the politicization of twentieth-century history has perpetuated linear and homogeneous explanations of the nascence of women's public activism. Feminism has been used as a symbol of progress with which intellectuals constructed the history of the nation. Too often, political historians have connected women's agency to liberal feminism and colonial nationalism, whereas labor historians have portrayed women's activities as linked to class struggle. The history of women's legal and social emancipation in early twentieth-century Korea has thus been irrevocably tied to the crusade for national liberation.[7]

Although the precise origins of the women's rights movement in Korea remain contested, much of the scholarship on modern women accredits the rise of female social and political agency to indigenous impulses. According to civic activist and attorney Yi T'aeyŏng, the real advance toward the emancipation of women began "only after the Liberation in 1945." Social scientists likewise maintain that Japan's defeat on August 15, 1945, not only meant release from foreign domination but signaled the "revolutionary moment" in which the emancipation of women in Korea took on more concrete form. Such analyses rightly highlight the reality that, between 1910

and 1945, large-scale organization for women's legal and economic autonomy was hindered by colonialism. Electoral suffrage and wide-ranging
drives for equal rights for women on legal, political, economic, and civic
grounds advanced only after the enactment of the Constitution in 1948.[8]
While Yi defines women's rights in terms of suffrage, historians including
Park Yong-ock locate protofeminist thought in the revisionist campaigns of
the eighteenth and nineteenth centuries. They trace the birth of Korean
feminism to the initiatives of Korean intellectuals, such as the development
of Practical Learning (Sirhak), the Enlightenment movement (Kaehwa undong) of the 1880s, the Tonghak (Eastern Learning) uprising of 1894, and
the 1896 formation of the Independence Club (Tongnip hyŏphoe).

While legal suffrage took longer, large-scale efforts for the expansion of
women's social and political rights first emerged in the late nineteenth century. Ideas of women's liberation were encouraged by female missionaries
from Europe and the United States, arriving in the 1880s after the diplomatic opening of Korea. Notwithstanding that Protestantism was imported,
church operations quickly indigenized and Korean Christians soon took on
evangelical responsibilities. Local Bible women (chŏndobuin) aided popular
literacy by teaching the vernacular script (han'gŭl) that opened opportunities for rural women.[9] Associated with religious service was health care,
which also broadened the range of female influence. The Women's Foreign
Missionary Society established the first women's hospital in Korea in 1887
on the compound of the newly built Ewha Woman's School (Ewha yŏhak-
tang). As with teaching and proselytizing, Korean professionals quickly took
over medical positions. A student of Ewha, Esther Kim, became the first female Korean physician, receiving her degree from the Woman's Medical
College of Baltimore in 1900.[10] Although a few attended women's medical
schools abroad, most remained in Korea and became nurses. Expanding
their traditional roles of womanhood as nurturers and healers, female evan-
gelists, nurses, physicians, and teachers realigned the limits of domesticity
and women's work from the home to the community.

Just as twentieth-century scholars have emphasized the indigenization of
Protestantism, the origins of Korean feminism have been coupled with na-
tionalism. Therefore, historians often maintain that the ideological bases of
Korean feminism can be traced to the designs of Sŏ Chae-p'il and Yun
Ch'i-ho, who among others founded the Independence Club in 1896.

Moved by Protestantism, Western education, and notions of liberal reform, the club aimed to "[tap into] a hitherto untapped resource" through women's education. One of the first women's rights organizations in Korea, the Praise and Encouragement Society (Ch'anyanghoe), joined the club in its support of modern women's education. With more than four hundred members, the Praise and Encouragement Society established the first nonreligious women's educational institute in Korea, the Sunsŏng Girls' School.[11]

According to liberal feminists, associations like the Praise and Encouragement Society were the first revolutionary women's organizations. Their ideological leanings, however, sustained economic and cultural elitism. Largely composed of the privileged, such organizations upheld the power of women in high society whose sense of *noblesse oblige* prompted them to educate their less-fortunate sisters. As the 1906 prospectus of a Ladies Commercial Association conveys, although women pursued new rights, they did so through maternal and domestic means, uniting their *"inferior* knowledge, strength, wealth and judgment."[12] Mediated by the patriarchal order espoused by both Confucianism and Protestantism, advances in women's education were often superficial. In practice, women's schools served as secondary homes and female teachers often paid more attention to the welfare and upbringing of future wives-to-be than to academic instruction.[13] Following the arrangement of marriages by Methodist leaders, the Ewha commencement of 1908, for instance, became a graduate wedding ceremony. Women's organizations in Korea multiplied in numbers during the first decades of the twentieth century, but their perspectives and procedures retained residues of conservatism.

Feminine activism before 1910 was championed by shifting, ideologically fluid, issue-focused coalitions. After annexation, sectional interests and divisions in society were minimized if not altogether suppressed. Stirred by the spirit of renewal, female reformers became strategic partners in the colonial nationalist effort. The ideological shift sparked by annexation is best described by an anonymous female student who called for an end to the "old-fashioned" model of "good wife and wise mother," alleging that education should attempt to develop self-reliant persons "who could serve family, society and humanity at large."[14] Nationalist women's associations, such as the Patriotic Women's Society (Aeguk puinhoe) and the Women's Society for Korean Independence (Taehan tongnip puinhoe), merging under the name of the Patriotic Women's Society in 1920, advocated suffrage for Korean men and women.[15] Formed seven years later, the largest female organization of the colonial era, the Kǔnuhoe (Friends of the Rose of Sharon), offered a united forum for nationalists and socialists. Prominent intellectuals of the day, such as Cho Wǒnsuk, Kim P'ilsu, Yi Hyǒn'gyǒng, and Kang Chǒnghi, organized communitarian cells in the Kǔnuhoe. In addition, socialist leaders including Hǒ Chǒngsuk, Paek Sinae, Pak Hojin, Chǒng Chǒngmyǒng, U Pongun, and Sim Sasuk, joined its ranks. But the society's leadership roles were occupied by better-known nationalist leaders such as Helen Kim (Kim Hwallan) and Yu Okgyǒm. Its literary organ, the *Kǔnu*, ran a selection of editorials concerning the plight of working women as well as gender and the economy, but Kǔnuhoe leaders failed to form a unified plan of action.[16] Because the Kǔnuhoe was a corollary of the New Korea Society (Sin'ganhoe), members of the Kǔnuhoe also assumed that national liberation was the primary concern of women and workers alike.

Emerging in the late 1920s, united front organizations claimed to welcome diversity but ultimately sought to amalgamate the nation's left-of-center political activism in the face of inevitable bifurcation. A year after the establishment of the Kǔnuhoe, thirty socialist women founded the party's first female organ, the Socialist League of Korean Women (Chosǒn yǒsǒng tonguhoe). Inaugurated under the supervision of the imperial police and fifty male leaders, the league promoted feminism as a logical corollary of socialism. Still, socialism was not an inherent ally of feminism in Korea. Just as an "ideological schism" formed within the general nationalist movement, with those upholding social reconstruction (*sahoe kaejo*) on the one hand and those endorsing the commonwealth (*kongje*) on the other, feminist groups likewise split into reformist and radical factions.[17]

Works in the English language by Protestant leaders such as Helen Kim and Louise Yim (Im Yǒngsin) place women's liberation second to national liberation, but opposing ideas also prevailed. Increasingly, women contributed to the new vernacular press, burgeoning in the 1920s. Women novelists published fiction serially in newspapers including the *East Asia Daily* (Tonga ilbo) and in journals such as *New Woman* (Sinyǒsǒng), *Woman's World* (Yǒjagye), and *New Family* (Sin'gajǒk). Writers such as Kim Wǒnju and Na Hyesǒk diverged from traditionally feminine themes by forming the literary journal *Ruin* (P'yehǒ). Radical proposals for social reform appeared for the first time in the 1920s as authors including Kim Wŏnju, in her novel, *Awakening* (Chagak), suggested that women's liberation necessitated the abolition of the family system. Also departing from the conventions of domesticity, Kang Kyŏngae exposed the lesser-known details of women's experiences in the factories in the *Problems of Humanity* (In'gan munje). Despite these critical voices, the most popular topics of discussion in women's journals and newspapers focused on women's customary roles in tales of romantic love. A well-known example was *I Am Loving* (Nanŭn sarang handa) by Kim Myŏngsun. In the end, colonial campaigns for women's liberation allowed for the publication of dissenting ideas but aimed for the *gradual* attainment of equal rights for women.[18]

While nationalism was espoused by intellectuals in late nineteenth-century Korea, colonial nationalism, fueled by anti-Japanese sentiments, emerged during the first decade of "military rule" between 1910 and 1920. As a response to the overwhelming resistance displayed in the March First demonstrations, which erupted throughout 1919 calling for Korean independence, the Government-General adopted a policy of "cultural rule" on the peninsula, which enabled patriotic groups, journals, vernacular periodicals, and social initiatives, including those calling for women's rights, to flourish throughout the 1920s. Official tolerance, however, waned during the 1930s, ending altogether by the start of the Sino-Japanese War in 1937. By revoking the liberal promise of citizenship while simultaneously endorsing a rational order, colonization debunked the ideological foundations of modernity.[19] Therefore, Korean nationalism challenged the duplicity of imperialist logic. Unwittingly, however, colonial nationalists also reproduced the same hierarchies introduced by the colonists in imposing cultural solidarity on an otherwise heterogeneous people. They presumed the impotence of the masses, epitomized by the status of colonial women and children, and their need for representation. As Japanese colonialism nullified the political authority of natives, patriotic intellectuals, speaking for the entire ethnic nation, muffled the myriad of voices in Korea.

Despite the scope of resistance activities throughout the colonial era, the nationalist effort failed to turn ordinary Koreans into anti-Japanese revolutionaries. Colonial nationalists opposed Japan's political domination but, by hewing to the social and economic standards of Japanese elites, they reinforced many aspects of Japanese cultural hegemony.[20] Because of their com-

plicit tendencies, late twentieth-century intellectuals have referred to many
colonial elites as "collaborators."[21] In most cases, however, it is not clear
whether Korean intellectuals supported Japanese colonialism exclusively or
complied with the occupiers to further Korean modernism. Throughout this
work, I argue that modernization and colonization evolved concurrently in
early twentieth-century Korea. Just as the procedures of indigenous and foreign development affected each other, so too did peoples' outlooks, identities,
and cultures. Postcolonial scholars have termed this merging of the colonizers' and the colonized cultures as hybridization, or the fusion of two influ-
ences. Not just colonial, hybridization was also a product of modernization.
Modernization—or the concomitant processes of colonization, capitalization,
industrialization, urbanization, and rationalization—transformed the settings
of life and work and altered the characteristics of human alliances, making
them more malleable and multifaceted than premodern ties.[22] Accentuating
the cultural significance of nationalist initiatives elides the variations in individual actions and outlooks that prevailed in early twentieth-century Korea.

## Socialism and Feminism

In the south, the rise of Korean feminism is attributed to cultural nationalism, whereas in the north, it is credited to state socialism. Scholars to the left
of the political spectrum, including Kang Tongjin and Kobayashi Hideo,[23]
contend that women's rights initiatives were sparked by colonial capitalism
and the proliferation of communist and socialist organizations, which inspired workers' activism in the late 1920s and the early 1930s.[24] They maintain that the creation of the League of Korean Workers (Chosŏn nodong
ch'ongdongmaeng) in 1924 and the constitution of the Korean Communist
Party (Chosŏn kongsandang) in 1925 evidence the formation of a "working
class." Kobayashi asserts that communists and labor activists found support
among workers in coal mining, construction, and transport as well as in
chemical enterprises, including footwear, rubber, and nitrogen fertilizer
(chilso piryo) production. Because reserves of natural resources such as coal
and iron ore were located chiefly in the P'yŏngan and Hamhŭng provinces,
most heavy industrial plants were situated in the north. Therefore, many
historians of colonial Korea have assumed that the early twentieth-century

labor movement relied on the activities of male workers in heavy industrial plants based in the northern provinces.[25]

Pointing to the escalation of large-scale uprisings such as the Wŏnsan general strike in 1929 and the 1930 Pyongyang rubber workers' general strike, these scholars allege that the late 1920s and early 1930s gave birth to a new generation of more militant activists. As noted by Kobayashi, the significance of communist infiltration was particularly visible in the reorganization of unions in metropolitan centers such as Pyongyang, Sinŭiju, and Hŭngnam. The formation of the Sinŭiju Factory Laborers' Union (Sinŭiju kongjang nodong chohap) in October 1930 and the Red Hŭngnam Chemical Workers' Association (Hŭngnam chŏksaek hwahak nodong chohap), as a division of the Hamhŭng Red Labor Alliance of Korea (Chosŏn chŏksaek nodong chohap Hamhŭng wiwŏnhoe) in 1931, demonstrated this new leftist attention to the heavy industrial constituency.[26]

Despite their emphasis on the activism of working men, socialists and communists also formed organizations for the mobilization of women and youth. The Chosŏn Women's Youth Association (Chosŏn yŏja ch'ŏngnyŏnhoe) and the Chosŏn Women Comrades Society (Chosŏn yŏsŏng tonguhoe), although allied to larger associations such as the Korean Workers Mutual Aid Society (Chosŏn nodong kongjehoe) as well as the Korean Worker-Peasant League (Chosŏn nodong ch'ongdongmaeng), pursued the "exclusively feminist goal of women's liberation," and the expansion of basic individual rights, economic self-sufficiency as well as "national independence." Societies such as the Love Native Products Society (T'osan aeyonghoe) and the Chosŏn Women's Cooperative Society "propagandized the spirit of the women's movement and took the problems of the working woman as their main issue." Throughout 1924, the Chosŏn Women's Cooperative Society established over forty branches throughout the peninsula and sponsored forums for feminist, nationalist, and proletarian discussions. By 1925, leftist women's coalitions collaborated with larger umbrella organizations such as the Korean Communist Party (Chosŏn kongsandang) and the Korean Communist Youth Association (Koryŏ kongsan ch'ŏngnyŏnhoe).[27]

The 1933 Platform of the South Chŏlla provincial branch of the Communist Party of Korea (Chosŏn kongsandang) illustrates the extent of leftist attention to female factory workers. Apart from calling for the abolition of the dormitory system (*kisuksa chedo*) and the production-based wage system (*togŭp chedo*), party members also opposed the differentiation in wages between Japanese and Korean employees, as well as the exploitation of women, children, and Chinese workers. In addition, they proposed the adoption of an eight-hour workday for adults, a six-hour day for youths under eighteen years of age, and a four-hour workday for those under sixteen. They also recommended the abolition of work for children younger than fourteen and the termination of the child labor system. The communists of South Chŏlla province appealed for a "forty-six hour work week and a paid holiday once a week," as well as gender-specific benefits such as maternity leave and breaks for breast-feeding.[28] Although the Red labor and Red peasants' movements influenced some regions including the Chŏlla provinces, communist groups were not able to attain wider levels of organization.[29] Since the clandestine formation of the first Communist Party in 1925, four parties rose and fell. Provincial and local communist affiliations gained constituents among workers and peasants, but "a close relationship between the laborers and peasants and their leaders did not exist." Ultimately, communists failed to "reach out" to the common populace.[30]

In South Korea, although lower-class women's consciousness and activism are often linked to the labor organizations of the 1970s and 1980s, historians and sociologists such as Yi Hyojae, Kang Isu, and Yi Chongok[31] have unearthed earlier examples of female wage work, social consciousness, and labor activism. Their scholarship exposes the details of Korean women's first entry into factory labor in the late 1910s and 1920s, specializing in light industries such as silk reeling, cotton spinning and weaving, as well as rice and food processing.[32] Nevertheless, historical portrayals of colonial factory women have been Janus-faced; while factory workers of the colonial era have been vaunted as the first wage-earning women in Korea, they have also been viewed as victims of an oppressive, imperialist-led capitalization (*chabon chuŭihwa*) process. Women workers of the colonial era expressed their social and political consciousness by demonstrating in the thousands. But, as Yi Hyojae and others conclude, their actions ultimately called for the liberation of the nation.[33] Colonial Korean women's sojourns in factories and cities have thus been depicted as transitory and filled with oppression, struggle, and suffering. The resistance of women workers is underscored for its patriotic value, but many scholars maintain that colonial women were unable to affect the level of social and economic change necessary for the greater liberation of women in the public sphere.

While leftists posed formidable plans on paper and a threat to colonial authorities, most women's labor demonstrations were largely unaffected by socialist praxis. Rather, factory women's protests surfaced specifically, among workers in the same industry or in the same region. Apart from the boycotts of hosiery workers, which were concentrated in a few areas including South P'yŏngan province, strikes among rubber workers, silk reelers, and cotton textile manufacturers erupted throughout the Korean peninsula. Rubber shoemakers and the employees of large-scale spinning and weaving factories might have been encouraged by union direction or communist infiltration. But the vast majority of strikes by women, especially among cotton textile and silk-reeling workers in small to medium-sized mills, were not directly mediated by such organizations.[34] The culturally liberal policies of the 1920s and the material crises of the late 1920s and early 1930s brought on the formation of public interest groups of nationalists, socialists, and feminists, to name a few. These groups served as important legacies for late twentieth-century activists but were, for most contemporary factory women, movements without their membership.

As described by E. P. Thompson, the "notion of class entails the notion of historical relationship," and classes form when people, as a result of common experiences, "feel and articulate the identity of their interests as between themselves," and as against other persons "whose interests are different from [and usually opposed to] theirs." Class consciousness, therefore, is the manner in which these experiences are handled in cultural terms: "embodied in traditions, value-systems," ideas, and institutions. Although *class* can refer to fundamental economic dissidence between owners and producers, *classes* can be conceptualized as social and political groupings.[35] Following Thompson, I refer to class as a very loosely defined body of people who share similar interests, social experiences, traditions, and value systems; people "who have a *disposition* to *behave* as a class, to define themselves in their actions and in their consciousness in relation to other groups of people in class ways."[36] Thus, class is an *occurrence* that is relative to time, space, and experience.

Although both male and female workers in early twentieth-century Korea assembled for proletarian interests, the absence of the rubric of the "working class" in documents and testimonies, as well as the specificity of workers' alliances, suggests that women workers did not identify themselves as members of a distinct class. Moreover, that gender norms bound women to certain forms of labor and lifestyle invariably made women's working-class consciousness dissimilar from that of men. Because the notion of class is used without a clearly specified definition, "debates about class often become conversations in which people talk past each other because they are talking about different dimensions of class."[37] While it is arguable that colonial working women composed a class of their own, the deviations between their tasks, experiences, and inclinations indicates that to confine women workers to one class would undermine their more intimate modes of identification.

In a time when traditional ideas about class and its meanings were evolving, colonial Korean factory women held to different identities than those exclusively affiliated with the "working class." Although some mention of class (*kyegŭp*) is made in extant sources, the language of labor (*nodong*) articulated contemporary debates on women's work. Factory women were usually born into the lower classes, composed of tenants and wage laborers, but because of the transitional nature of the colonial economy, most likely they did not conceive themselves as part of an "industrial working class." Their emphasis on labor, as determined by skill and efficiency rather than class, can be seen by the predominant oral and written use of the term *worker* (*nodongja* or *kŭlloja*).[38] Workers' alliances with political associations were not mutually exclusive. Factory women often affiliated themselves with Protestant congregations,[39] as well as nationalist organizations, without compromising their professional or familial connections.[40] While the sheer size of the labor force in the 1930s offered an unprecedented constituency for partisan groups, this magnitude also allowed for greater factionalism. The vast number of peasant uprisings and labor strikes during the 1920s and 1930s attest to the political consciousness of lower-class women, but whether their allegiances were permanent or transient is difficult to measure. Most likely, during times of recession and labor surplus, women united to safeguard their interests, but in times of labor shortage, women sought to ascend the ladders of the labor hierarchy independently. Not affixed by the narrower confines of class or ethnicity, the activities and identities of working women were not just heterogeneous but also contextually contingent and, therefore, constantly changing.[41]

*Reexamining Women's Political Consciousness*

Without doubt, efforts for women's rights in colonial Korea drew their inspiration from liberalism and its nineteenth-century legacy of suffragism. Armed with emancipatory brands of Protestant thought, suffragists in Europe and North America strove to promote women's authority by enlarging their traditional roles. The Korean contributors to the 1884 founding of the Ewha Woman's University embraced the ideals of transatlantic suffragists, even adopting the English neologism "wom*an*," which symbolized the unity of the female sex and proposed that all women have "one cause, one movement." By the 1910s and 1920s, activists increasingly espoused European feminism, signaling a new phase in the debate and agitation concerning women's rights and freedoms. Because feminism, particularly the liberal feminism extolled by authors such as Ellen Key, did not rely on suffrage as a unifying goal, it was attractive to Korean women who pursued social influence under colonial governance. Although seeking separation from the former "woman movement," Korean feminists, like feminists elsewhere, still carried vital connections to transatlantic suffragists, who aspired to extend the boundaries of the domestic sphere rather than break through them altogether.[42]

Viewing feminism as a cultural response to patriarchy, Gerda Lerner defines feminist consciousness as the "awareness of women that they belong to a subordinate group"; that they suffered wrongs as a group; that their subordination is not natural but socially determined; that they must join other women to remedy these wrongs; and finally, that "they must and can provide an alternate vision of societal organization in which women as well as men can enjoy autonomy and self-determination."[43] Elite women of colonial Korea might have held "feminist consciousness," but to attribute political awareness exclusively to conspicuous historical actors distorts the complexities of everyday identities. Despite the theoretical intricacies involved in exploring the differences between the ideals of elite and ordinary women, to understand the greater evolution of female agency in modern Korea, a broader definition of women's political consciousness is needed.

Proliferating in the late twentieth century, feminist theory relied on overarching categories that ascribe the origins of sexual hierarchy to either political or economic foundations. Authors such as Kate Millett and

Shulamith Firestone trace social asymmetry to the discovery of patriarchy, arguing that "the appropriation by men of women's sexual reproductive capacity occurred prior to the formation of private property and class society."[44] In contrast, Juliet Mitchell, Claude Meillassoux, Leopoldina Fortunati, and others extend Friedrich Engels's thesis that the agricultural economy impelled sexual divisions of labor, maintaining that gender inequalities were determined by modes of production.[45] The debate over the origins of patriarchy had implications for women's consciousness and action: If patriarchy resulted from economic structures, then efforts for women's emancipation should be allied with that of class. If, on the other hand, patriarchy preceded systems of private property, the origin of social inequality would be found in sexual politics or the creation and continuation of gender norms.

These debates among feminists shook the viability of "womanhood" and, by extension, other features of social distinction such as class and ethnicity. As Judith Butler explains, feminism had assumed some "existing identity, understood through the category of women," who not only initiated feminist interests and goals within discourse but constituted the subject of political representation. Identity categories, such as sex, were always normative as well as descriptive, and as such, exclusionary. Christina Crosby extends the analysis of gender and subjectivity to individual cases and posits that personal identities were defined by distinctions. She asserts that differences, which seemed to refract and undo a substantive identity, actually reflected a multifaceted, modified but "all-too-recognizable subject."[46] These authors imply that scholarship on women has thus been restricted by the very structure of power through which emancipation is sought. The irreducibility of identity, as outlined by poststructuralist feminist theorists, indicated that gender was no more of a unifying characteristic than class or nationality.

Nonetheless, by clarifying the relations between gender and production, feminists challenged the dichotomization of the personal and political spheres of life. Feminists of the 1960s recognized that electoral suffrage and other forms of action within the prescribed, male-dominated bounds of "politics" were woefully insufficient. Because women lived intimately with their "oppressors," in isolation from each other, women were unable to see their "personal suffering as a political condition."[47] Although male-female relations surfaced as interplays between two isolated individuals, they were in fact driven by culturally embedded paradigms of power. Because social structures were products of patriarchy, female intellectuals had to regard personal experience as the basis for analyzing common situations. Late twentieth-century feminists argued that political categories cannot be limited to the realms of governance but must be extended to all social relations[48] through their declaration that "the personal is political."[49]

While numerous feminists attribute the dichotomization of the personal and the political to the incipience of patriarchy, others including Eli Zaretsky locate its origins in capitalism. According to Zaretsky, the division between the labor in a capitalist economy and the private labor of women in the home was closely related to a second division: "between our personal lives and our place within the social division of labor." With mechanization, work and life became separated; "proletarianization split off the outer world of alienated labor from an inner world of personal feeling." Modern corporations, including those in colonial Korea, depended on individualism: It was the individual who was hired, evaluated, paid, and fired. Thus, despite their partition, political order relied on the personal. The individual, however, was not isolated; he or she was connected to others in a family, in what Peter Berger describes as an institution that balances the anonymous aspects of personal autonomy with communal solidarity. Families functioned as individual gateways to society, often expanding an individual's spectrum of choices or agency through material or symbolic inheritance.[50] In modern capitalist societies, families were more than economic prerogatives; they were political units, legitimated by state support.

The social standings of female factory workers, as colonial subjects, as members of the lower classes, and as young women, were undoubtedly subaltern. Nevertheless, did factory women conceive of themselves as multiply subordinated? By declaring that knowledge is not founded on ontology, Donna Haraway implies that one's cultural characteristics do not necessarily translate into identity. Rather, self-knowledge was intimately connected to an individual's "shifting, mobile, simultaneous, multiple," and critical positions within his or her immediate communities and families.[51] Although their identities and behaviors often diverged, the activities and testimonies of factory workers demonstrate the prevalence of popular forms of gender, labor, social, and political consciousness among ordinary women in colonial Korea.[52]

## Methodology and Presentation

Few intellectual initiatives inspired historiographical reevaluation to the extent of those emerging in the late-twentieth century. Feminist revisions of labor historiography, for example, have broadened understandings of women's productive roles in premodern and modern societies. Similarly, labor historians have called for approaches that "Thompson himself would probably applaud: the integration of working class history, in all its aspects, with a larger history of society."[53] Shifting away from studies of unions and ideologues, scholars increasingly focus on the collectivity of laborers with common roots in the countryside and common experiences. Postcolonial theorists, such as Partha Chatterjee, Gayatri Spivak, and Homi Bhabha also challenge capital's primacy by presenting the paramountcy of community. According to Chatterjee, because European notions of citizenship did not apply to colonial subjects, colonized people constructed their national identities "within a different narrative." Shifting their focus from the nation to the community, postcolonial intellectuals call for the unearthing of hybrid forms of ideological translation and identity formation.[54] In this spirit, rather than "culture *imposed* on the popular classes," this book focuses on "culture *produced* by the popular classes." But taking the lower classes as producers of their own cultures raises a succession of questions: "To what degree is the first, in fact, subordinate to the second? And, in what measure does lower class culture express a partially independent content? Is it possible to speak of reciprocal movement between the two levels of culture?"[55]

My methods for reconstruing the lives and labors of colonial Korean women draw insights from recent scholarship and, in particular, the findings of late twentieth-century social theorists. Although historical interpretation remains susceptible to the distortions of representation, I have pursued accountability and responsibility in my analysis. To explore the subjectivities of female industrial labor, I connect the rise of women's labor with the surrounding character of economic development and larger social progressions, such as the growth of wage work in the manufacturing and agricultural sectors, as well as patterns of migration. Historical, cultural, and regional distinctions, including the climates of colonial modernization and wartime mobilization, help make comprehensible the dynamics between economy and society. The character and composition of the industrial labor force evolved considerably throughout the colonial era. Although the earliest factory workers in Korea worked on state-sponsored projects in the 1900s, private firms began recruiting rural youth for labor soon after. Examples of some of the first factory girls were the eighteen workers who in November 1911 activated the production lines of the Kyŏngsŏng Cord Company that became one of the largest textile enterprises in twentieth-century Korea.[56]

Though purportedly scant, materials on colonial factory women are plentiful, even if disarrayed. The *Scrapbook of Newspaper Clippings* (Sinmun chŏlbal), compiled by the colonial government and archived in the Imperial University (Teikoku daigaku) collections of Seoul National University Library, offers a quick reference to articles and editorials on the activities of laborers, peasants, and students. Recruitment records and other sources found at Yongin archives of the Kyŏngsŏng Spinning and Weaving Company also shed light on the varieties of labor mobilization.[57] The *East Asia Daily* (Tonga ilbo), *Chosŏn Daily* (Chosŏn ilbo), and local newspapers depict events in everyday life. Records of the Government-General of Korea (Chōsen sōtokufu) including statistical yearbooks (*tōkei nenpō*), monthly research reports (*chōsa geppō*), and the national census (Chōsen kokusei chōsa hōkoku) furnish detailed and relatively reliable data on industries, factories, and workers from 1910 to 1945.[58] Although published less frequently, the censuses are far more comprehensive, including accounts for household industries lacking in the statistical yearbooks. The first nationwide population census was taken in 1925 and was repeated every five years thereafter, with the last census taken in 1944. Periodicals of the colonial era, including *New Woman* (Sinyŏsŏng), *New Family* (Sin'gajŏng), *New East Asia* (Sintonga), and *Creation* (Kaebyŏk), also provide a wealth of description that illustrates the issues and concerns confronting Korean culture and society in the 1920s and 1930s.

The destruction caused by the Second Sino-Japanese War (1937–45) and the Korean War (1950–53) meant the misplacement and ruination of historical records, and thus, tracing the experiences of the second generation of workers in colonial Korea through recorded material was no easy task. In spite of the decline of record keeping in the 1940s, the wartime reports of bureaus such as the Chōsen Society for Public Welfare as well as *Kyŏngsŏng Daily* (Keijōnippō) and *Daily News* (Maeil sinbo) articles sketched rough portraits of transformations in the labor market in the last decade of colonial rule. What was lacking in written sources, however, lived in the memories of elderly Koreans who survived colonization, civil war, and the rigors of late twentieth-century modernization. After contact with the Society for the Survivors and the Bereaved Families of the Pacific War (T'aep'yŏngyang chŏnjaeng hŭisaengja yujokhoe), I was given access to some of its records and located the whereabouts of several dozen women who worked in factories in the late 1930s and 1940s. Of fifty-three inquiries, seven responded, and their statements on factory labor during the last decades of the colonial era contribute to the rich mosaic of the lives and identifications of women who worked in early twentieth-century Korea.[59]

Eight former factory women and one male clerical employee are referred to in this study (Table 1.1). Seven of the nine accounts are taped interviews ranging from forty-five minutes to two hours that I conducted in 1999. Although the Society for the Survivors and the Bereaved Families of the Pacific War and other civil organizations had memberships in the thousands, the majority of those who worked in the 1930s were deceased by the late 1990s. Thus, the borrowed tapes of two interviews conducted in 1991 by Kang Isu of Sangji University offered invaluable data on the industrial conditions confronting those who worked several years earlier. Of the nine firsthand informants, six worked in textiles and three in steel manufacturing.[60] Kang's informants were Yi Chaeyun, a former piecer at the Kanegafuchi factory in Kwangju, and Kim Yŏngsŏn, a clerk who had retired from the Kyŏngsŏng factory in Yŏngdŭngp'o.[61] The remaining textile workers were: Han Kiyŏng, an employee of the Dai Nippon (Tae Ilbon) factory in Yŏngdŭngp'o between 1944 and 1945; Kang Pokjŏm, a piecer at the Tōyō (Tongyang) factory in Yŏngdŭngp'o from 1942 to 1945; Yi Chungnye, a carding operative at the Chongyŏn (Kanegafuchi) factory in Kwangju in 1944; and Kim Ŭnnye, who was a paid laborer in both a tobacco factory and a hosiery plant in Pyongyang during the late 1930s before being taken to Manchuria as a comfort woman in the 1940s. The recollections of those who worked in textile enterprises provided much information on the settings and experiences of colonial workers, but they resembled documented descriptions of factory women in the 1920s and furnished little insight

**TABLE 1.1**
*Interviews*

| Informant | Date of Interview | Year of Birth | Province of Birth | Company | City of Factory | Task in the Factory | Length of Employment | Occupation after 1945 |
|---|---|---|---|---|---|---|---|---|
| Han Kiyŏng | 6.11.1999 | 1930 | Kyŏnggi | Tae Ilbon (Dai Nippon) Textiles | Yŏngdŭngp'o | Piecing Operative | 12.1943–8.1945 | Housewife |
| Kang Pokchŏm | 6.3.1999 | 1925 (?) | North Ch'ungch'ŏng | Tongyang (Tōyō) Textiles | Yŏngdŭngp'o | Piecing Operative | 2.1942–8.1945 | Domestic Servant/ Cleaner |
| Kim Chŏngmin | 6.8.1999 | 1926 | North Chŏlla | Fujikoshi Steel | Toyama-shi, Japan | Dormitory Supervisor | 1.1943–8.1945 | Physician of Chinese Medicine |
| Kim Chŏngnam | 6.5.1999 | 1930 | North Chŏlla | Fujikoshi Steel | Toyama-shi, Japan | Lathe Operator | 3.1943–8.1945 | Teacher |
| Kim Ŭnnye** | 6.12.1999 | 1925 | South P'yŏngan | Pyongyang Hosiery/ Pyongyang Tobaccos | Pyongyang | Piecing Operative | 1936–1939 approx. | Various Forms of Wage Work |
| Kim Yŏngsŏn* (Male) | 6.27.1991 | 1922 (?) | South Kyŏngsang | Kyŏngsŏng Textiles | Yŏngdŭngp'o | Clerical Employee | 1938–8.1945 approx. | Clerk |
| Pak Sun'gŭm | 6.8.1999 | 1929 | North Chŏlla | Fujikoshi Steel | Toyama-shi, Japan | Lathe Operator | 12.1944–8.1945 | Farmer |
| Yi Chaeyun* | 10.4.1991 | 1931 (?) | South Chŏlla | Chongyŏn (Kanegafuchi) Textiles | Kwangju | Piecing Operative | 1944–8.1945 | Activist |
| Yi Chungnye | 6.3.1999 | 1928 | South Kyŏngsang | Chongyŏn (Kanegafuchi) Textiles | Kwangju | Piecing Operative | 2.1944–2.1944 | Various Forms of Wage Work |

Source: Taped Interviews ranging from 45 to 120 minutes.
* Interview conducted by Kang Isu of Sangji University.
** Also drafted as a comfort woman.

into the changes of the late 1930s. The testimonies of Kim Chŏngnam, Kim Chŏngmin, and Pak Sun'gŭm, who were employed in the machine and machine tools sector, however, revealed drastic transformations in the character and composition of the wartime workforce. The stories of these women, recruited through wartime labor mobilization projects, demonstrate how some women not only crossed industrial boundaries but acquired skilled positions. Obviously, the testimonies of nine individuals cannot be considered to be representative. But, by fostering more critical and creative analyses of documentary evidence, they offer rare perspectives into the more personal meanings of labor.

Although the testimonies of workers of the late colonial era contribute greatly to the interpretation of the second generation of laborers, this project surveys the lives and labors of female factory workers *throughout* the colonial era. Before factory women are examined in depth, however, I situate colonial Koreans in space and time and describe the environments which engendered female industrial labor in Chapter 1. Chapter 2 provides an overview of the peninsula's historical and geopolitical background and outlines the progressions of colonial economic development and the emergence of women's wage work from the late 1910s to the 1930s. While the chapter attempts to furnish broad numerical and proportional sketches of colonial workers in general and wage-earning women in particular, discrepancies in and between the Government-General sources made more sophisticated analysis of the statistics untenable. Chapter 3, in contrast, recovers glimpses of life inside the factories by illustrating the terms and conditions of labor in the cotton spinning and weaving, silk-reeling, and rubber shoemaking mills. By reviewing the collective actions of workers and their modes of association, Chapter 4 demonstrates some of the varieties and extents of working women's political consciousness. Chapter 5 examines the war and its impact on women workers. The first half of the chapter outlines the operational changes in the procedures of production and labor mobilization during the late 1930s and women's roles within Japan's designs for victory. The second half of the chapter, however, links the transformations of the last decade of colonial rule with the stories of firsthand observers. Despite limitations posed by scant and incomplete written records and the diminishing availability of eyewitnesses, extant information on wartime workers clearly indicates that historical time and circumstance helped shape the working lives of Korean women. Through their experiences of labor, the values derived from the family were reinforced or discarded, but women also embraced new professional statuses and identities. The second generation of colonial workers—the *war generation*—diverged from the models set by their families and communities through their life courses. But a closer reading of their postwar histories reveals that many, by providing for their children and continuing the family line, ultimately fulfilled their parents' ambitions. Although the conclusion elaborates on the legacies of colonial women workers, it is not intended to examine women and work in the post-Liberation Koreas thoroughly, because exploration into such a topic would demand, at the least, another book.

As in Europe and North America, the term *new woman* (*sinyŏsŏng* or *sinyŏja*) came into vogue in Korea during the 1920s. According to popular understandings, women's public consciousness burgeoned among select female ideologues in response to Japanese imperialism. Critical research on colonial workers, however, strongly suggests that nationalist intellectuals, glorified in historic memory, were not the exclusive guardians of political consciousness. First emerging in the late 1910s, wage-earning women in colonial Korea gained the understanding that their social and economic subordination was not only unnatural but surmountable. United with their colleagues, they strove to resolve their work-, time-, locale-, and gender-specific problems. Whether they protested collectively or individually, women workers overcame personal and political boundaries, inspiring successors. Their heterogeneous identities were not associated exclusively with their labors but with their familial and communal obligations. But, because the relations between individuals and societies were dialectical, their changing attitudes and behaviors enabled them to alter their surrounding families, communities, and societies in turn. A former factory worker, Kim Chŏngmin, when asked if she thought of her consciousness as similar to the identifications of famous women such as Kim Hwallan and Im Yŏngsik, replied, "They were our elders, the generation before."[62] Less known to Kim, however, were working counterparts of elite nationalist women, such as the women who participated in one of the first citywide labor demonstrations of the colonial era, the 1923 Rubber Workers' General Strike in Seoul.[63] Despite the tumultuousness of their historical, political, and personal surroundings, by aiding in the construction, deconstruction, and reconstruction of early twentieth-century Korean society en masse, factory women served as some of the first "new women" who began to alter traditional conceptions of gender, labor, and power in twentieth-century Korea.

# Locating Korean Factory Women in Time and Place

One of the most universal and enduring symbols of industrialization has been the "factory girl." Torn from her family, surveyed constantly, overworked, and malnourished, depictions of female factory workers underscored the human costs of mechanization. The May 1929 inaugural issue of *Kŭnu*, the official journal of the Kŭnuhoe (Friends of the Rose of Sharon), featured an article by Pak Hojin, "Survey of Women Workers." A provocative portion of the piece narrated the dismal working conditions of the Chosŏn Silk Company (Chosŏn kyŏnjik chusik hoesa).

> Under the head-splitting sounds of machines . . . their work demands
> following the paces of mechanical protagonists. . . . Not being able to sit for
> twelve hours, when they first entered the mill, their legs were stiff and
> swollen and they thought they would die. They only look forward to leaving
> the factory.[1]

Such renderings of Korean factory women uncannily resemble descriptions of workers elsewhere—for example, those delivered by an English Member of Parliament, who detailed in 1838:

> Amongst other things, I saw a cotton mill—a sight that froze my blood. The place was full of women, young, all of them, some large with child and obliged to stand twelve hours each day. The heat was excessive in some of the rooms, the stink pestiferous and in all an atmosphere of cotton flue. I nearly fainted.[2]

Interpretations of factory women in colonial Korea and elsewhere characterize the lives and sentiments of working women in tragic terms. When viewing them, contemporary elites asked, "How did these women live?" Observers often responded: "Live? Ask rather how they [kept] from dying."[3]

Although such portraits have drawn attention to women's contributions to industrialization, they have also perpetuated misunderstandings of women's work before and after modernization. They have diminished the varieties and extents of female contributions to both the household and the mechanized modes of production. Just as important, they rest on the constancy of modern economic development across cultures and times. These renderings seem to echo Marx's dictum that the more industrially developed country presents to the less-developed country an "image of its own future."[4]

For better or for worse, the rise of industrial capitalism under colonial governance transformed the gender divisions of labor in early twentieth-century Korea. As human and natural resources were drawn from agriculture to manufacturing and commerce, farm girls moved from the countryside to urban mills, serving as some of the country's first factory workers. Whereas in 1910, there were 8,203 Koreans of both sexes employed in factories, by 1922, the number of women alone rose to 9,849, composing 20.5 percent of the total workforce. By 1933, women numbered 33,282 and constituted 33.5 percent of the industrial workforce, forming over a third of the total labor force throughout the 1930s and during the Pacific War (1931–45).[5]

According to pervasive views of working women in colonial Korea, female employment in the factories was largely circumscribed within the bounds of the textile industry.[6] So closely have they linked women's modern wage work with spinning and weaving that the degrees and diversities of women's participation in household, local, and regional economies remain unknown. Although the numbers of women entering textile mills increased rapidly in the 1920s, this growth allegedly tapered throughout the late 1930s as small-scale light industries were overrun by larger, technology-intensive enterprises that relied more on male workers. Closer research, however, evidences a more heterogeneous pattern of development. The expansion of heavy industrial projects prompted the upsurge of production from the 1930s, but light and household enterprises also contributed to this growth. Following modernization theories, many labor historians assumed that the late colonial Korean economy prioritized heavy industrial output exclusively, disenfranchising women workers in light industries. But a detailed examination of women's labor throughout the Second Sino-Japanese War suggests that female employment and economic agency continued. Although women remained in spinning, weaving, and silk reeling mills, by the late 1930s, a pronounced labor shortage and the urgency of war brought women into the heavy industries for the first time.

The prototype of the "mill girl" might adequately depict the entry of women into commercial sectors in early mechanizing environments. It does not, however, accurately represent the ways in which women in later developing economies gained employment. In colonial Korea, where modernization was rapid, women's factory work was less bound to the textile industry. This chapter highlights the significance of cultural, economic, and political specificity in determining the characteristics of women's industrial labor. To fully appreciate the experiences of Korean women, I first survey female factory workers across historical and regional boundaries, pointing to the ranges of cultural difference. Second, I describe how these differences were shaped in part by the distinct paces and traits of each economy's transition into the industrial mode of production, which across regions exhibit less structural constancy than contingency. Among other factors, historical timing and colonial governance spurred Korean mechanization, shaping the female industrial workforce in unique ways. Third, I turn to the distinctions of Korea's colonial, accelerated modernization—a process that was as much politically as economically driven. Finally, I explain how these cultural, demographic, economic, and political forces configured the emergence and evolution of the female workforce in early twentieth-century Korea. Synchronized with the motions of markets, policies, and popular practices, the characteristics of female employment underwent innumerable changes throughout the thirty-five years of colonial rule. An evaluation of factory women in early twentieth-century Korea, when compared to the rise of working women elsewhere, strongly indicates that although women across cultures worked in factories, their experiences diverged greatly according to economic, social, and political contingencies as well as historical and regional distinctions.

## Factory Women Across Cultures

Examining how women entered the factories across historical and regional boundaries confirms that the policies and practices of earlier and later mechanizers differed dramatically. In England, where the modern mode of industrial production was first instituted in the eighteenth century, women began wage work in small workshops with rudimentary management and manufacturing standards. By the nineteenth century, however, corporations in New England and Japan attached dormitories to their mills and housed their workers for maximum surveillance. In the textile plants of Germany, however, systems of residence, established by the turn of the twentieth century, were not pervasive.[7] Although a few mills in French cities such as Roubais and Amiens adopted tenements, company housing was scarce, attesting to the resilience of rural traditions in France. Cross-cultural analysis suggests that employment conditions for men and women were determined by the period of mechanization, the rate of production, and the adaptability of regional customs.

While light industries offered women jobs, their participation in textile and other manufacturing enterprises was far from uniform. As Joan Scott and Louise Tilly reveal, though the mills of England and France hired women, disparities in the rates of growth and the characteristics of employment between the two nations were conspicuous. Women in both England and France first entered manufacturing mills from the mid- to late eighteenth century; by 1851, of wage-earning women in England, 45 percent engaged in manufacture. In 1866 France, however, only 27 percent of wage-earning women were employed in factories, whereas 40 percent engaged in agricultural wage work.[8] The tenacity of the French agrarian tradition throughout the nineteenth century often made the distinction between industrial and agricultural production artificial. In eighteenth-century England, females accounted for around 50 percent of the textile manufacturing labor force. By contrast, in nineteenth-century United States and Japan, rural girls eventually made up 70 to 80 percent of their textile labor forces. Apart from being larger in scale, the methods of mobilization and residence in American and Japanese companies deviated from the practices of European businesses. Whereas in England and France, families tended to move to metropolitan centers together, recruiters in the United States and Japan selected young women from the remote countryside, which owners felt decreased unrest and bolstered more rigorous systems of supervision. To secure unmarried workforces, American and Japanese companies, especially cotton textile firms, installed elaborate procedures of recruitment and housing. Rather than go under, manufacturers of cloth and consumer goods in later industrializing economies remained competitive by investing in technology and comprehensive management guidelines.[9]

Inspired by the Victorian "cult of true womanhood," New England companies, including the mills of Lowell, Massachusetts, instated paternalistic regulations that confined their workers. The boardinghouses[10] of the Middlesex Company in Lowell, for instance, required their inhabitants to maintain clean living environments, to practice proper hygiene, and to retire for the night by 10 P.M. New England women received visitors and ventured outside of the factory compounds.[11] Although the boardinghouses of New England textile factories offered accommodations to female employees of all ages, dormitory residence was not compulsory.[12]

In contrast, factory women in East Asia endured greater degrees of physical confinement and restricted movement. By the turn of the twentieth century, the majority of female employees in Japan were recruited and housed by their employers, being allowed to leave factory premises once a year during the lunar New Year or the midsummer Obon festival.[13] Although early twentieth-century Japanese firms established corresponding recruitment and management procedures in their factories in China and Korea, the labor forces of the two countries evolved in markedly dissimilar manners. Instead of company-based recruitment and employment, a distinct "contract labor system" prevailed in Shanghai. Adhering to a local tradition of labor mobilization, private contractors acted as the recruiters and guardians of girls "bought" from the countryside. Contractors, often affiliated with the Green Gang, the nexus of Shanghai's criminal underworld, kept most of the workers' wages, providing meager housing and inadequate provisions in return.[14] In North China, more women were recruited and housed than in Shanghai. But, as Gail Hershatter notes of contemporary Tianjin, dormitory residents composed only a quarter of the industrial labor force as employees continued to evade the housing requirements of Japanese companies.[15] "Number ones," or foremen, selected women for work and allocated salaries in Shanghai mills, whereas in the factories of colonial Korea, corporate direction over recruitment and wages was strict. Although some of the factory conditions in Korea during the 1920s might have resembled Chinese settings, by the 1930s, managerial oversight strengthened and bore greater similarities to contemporary Japanese practices. The majority of Korean women in cotton spinning and weaving and silk-reeling enterprises confronted the systems of work and life that drew upon the traditions of Meiji (1868–1912) industrialization.[16]

Stark disparities in the standards of factory work in Korea and China in the early twentieth century highlight the significance of historical context. Although both countries experienced a period of accelerated growth in manufacturing following the World War I (WWI) economic boom, their development was shaped by their unique political and cultural climates. Both nations first industrialized under imperial hegemony, but China's state of semi-colonialism varied from the total colonialism prevalent in Korea. Local traditions, such as the contract labor system in Shanghai, integrated with the strategies of foreign businesses in Chinese cities. Older customs of recruitment existed in colonial Korea but were far less prevalent than the rationalized, Japanese system of corporate management. Colonial modernization was also affected by the evolving technologies of the colonizers. Whereas the colonial influences in China were North American, European, and Japanese, mechanization in Korea was guided almost exclusively by the Japanese. Overshadowed by local traditions, industrial development proceeded at a modest pace in early twentieth-century China, whereas modernization took a more accelerated form in contemporary Korea.

*Economic Development in Colonial Perspective*

According to nineteenth-century theorists, the transition from the agrarian to the mechanized mode of production was achieved through industrial capitalism, which required the prevalence of preconditions. Adam Smith referred to the need for "previous accumulation," whereas Karl Marx maintained that industrial infrastructures relied on the "original accumulation" of capital. Smith, Marx, and others believed that widespread agricultural reform, the rise of progressive elite classes, and the provision of social overhead capital were all "necessary preconditions" of industrial capitalism. Evaluating the differences between methods of modern economic development, Alexander Gerschenkron challenged these views by maintaining that while prerequisites existed in eighteenth-century England, they were absent in countries modernizing thereafter. Instead of preconditions, the state, banks, and ideology were the agents of mechanization. Although manufacturing emerged out of prerequisites in the first industrializing nation, in "moderately backward" areas, factories arose in cooperation with banks. Economists and historians have increasingly emphasized the discontinuities of development, positing that the more "backward" or later modernizing the country, "the more rapid will be its industrialization."[17] They propose that mechanization in eighteenth-century England was driven by capital accumulation. But, in nations modernizing in the nineteenth century, a closer association between banks and factories prevailed. In countries that first modernized in the twentieth century, the state, financial institutions, and businesses *simultaneously* promoted accelerated production. Dubbed by present-day economists as the "catch-up" theory, Gerschenkron's hypothesis of differentiated development is used to explain the rapid progress of economies that industrialized in the twentieth century.[18]

The timing and rate of Korea's economic development, however, was intricately tied to the balance of power and knowledge in East Asia. At the end of the nineteenth century, Korea stood precariously between the two empires of China and of Japan, "one in decline and the other ascendant."[19] Throughout the latter half of the Chosŏn dynasty (1392–1910), the Yi court held diplomatic and trade relations exclusively with China, prompting Europeans to refer to Korea as the "Hermit Kingdom." Korea's isolation, nonetheless, came to an end in 1876 with the Kanghwa Treaty, which initiated commercial and diplomatic relations with Japan. Korea signed treaties with other nations—opening the ports of Inch'ŏn, Pusan, and Wŏnsan to the United States in 1882; to England in 1883; to Germany, to Italy, and to Russia in 1884; and to France in 1886. By the turn of the twentieth century, the contest for power over Korea intensified between the Chinese, the Japanese, and the Russians. Carefully employing the civil unrest caused by the anti-foreign *Tonghak* (Eastern Learning) rebellion in 1894, the Chinese and the Japanese waged a war over Korea, both sides bringing troops to the peninsula to contain the rebellion.[20] The Sino-Japanese War (1894–95) and the subsequent, Japanese-driven *Kabo* reforms (Kabo kyŏngjang) established Japan's new role in Korean politics and set the economic and political foundations for eventual colonization. With the acquisition of Taiwan and the Liaodong Peninsula under the 1895 Treaty of Shimonoseki, Japan's new pan-Asian influence, as well as Korea's future role within its empire, solidified.

The inclusion of Hokkaido in 1869, Okinawa in 1871, and Taiwan in 1895 into the Meiji empire "laid the groundwork for later imperialist expansion."[21] Nevertheless, Russian involvement in the Triple Intervention that rescinded Japan's rights to the Liaodong Peninsula, their fortification of troops in Manchuria after the 1900 Boxer Uprising in China, and their entry into Korea with the 1903 opening of the Port of Yongamp'o threatened Japan's plans for the peninsula. Thus, in 1904, Japanese soldiers carried out a surprise attack on the Russian installations at Port Arthur. With Japan's victory in the ensuing Russo-Japanese War (1904–5), it acquired the southern half of Sakhalin. On November 17, 1905, it also enacted the Protectorate Treaty, which placed a Japanese resident-general (*tōkan*) in Korea. Five years later, an agreement in which the Korean emperor ceded all rights of sovereignty to the Japanese emperor was concluded. Formally announced by Emperor Sunjong on August 29, 1910, the Treaty of Annexation marked the end of the 518-year-old reign of the Chosŏn dynasty.

The waning of Chosŏn sovereignty coincided with the waxing of Meiji imperialism or the process of Japanese nation-building in the late nineteenth and the early twentieth centuries. The Meiji empire, encompassing territories previously uninhabited by the Japanese, reproduced "differentiation and inequality among people" it incorporated and "unformed," as it reformed existing structures and practices to manage and assert control over indigenous communities.[22] While colonized territories were incorporated into the Japanese empire, the rights and responsibilities of *naichijin* (Japanese or people of inner territories) varied dramatically from those of *gaichijin* (people of outer territories). Administering Korea as a Japanese colony meant "planting people" (*shokumin*) who would not only take over the government, economy, and society but spread "civilization." Korea, however, was rarely referred to as a "colony" (*shokuminchi*). Instead, Chōsen was called the "outer land" (*gaichi*) or the "peninsula" (*hantō*), a logic that was reinforced by defining the colonization of Korea as an act of "annexation" (*heigō*).

In the fall of 1910, the resident-general of Korea was replaced by a governor-general appointed by the emperor. The position was considered second only to that of the Japanese prime minister; until 1919, the governor-general of Korea was the only colonial chief responsible directly to the emperor, and then to the prime minister until 1942.[23] He resided over the Government-General of Korea (Chōsen sōtokufu), which consisted of the secretariat and seven departments: agriculture and forestry, business, education, finance, home affairs, justice, and police. The thirteen provinces were headed by provincial governors and councils. At the local levels were the cities (*fu*), counties (*gun*), islands (*shima*), towns (*yu*), and townships (*men*), which were administered by appointed officials. After roughly a decade or two of occupation, the peninsula became fully integrated into the Japanese empire, resembling the metropole in the workings of public education, health, "the military, telecommunications networks, museums, bureaucracy, fiscal policy, and, perhaps more tenaciously, in law courts, police and jails."[24] Through a myriad of bureaucratic and social scientific technologies, the colonial government set in motion the schemes to make the Korean population more amenable to political and productive programming. Just as Korean self-knowledge was mediated by Japanese knowledge about Korea, the ideals and behaviors of Koreans in the early twentieth century were affected by governmentality, that is, the "ensemble formed by the institutions, procedures, analyses and reflections, the calculations and tactics," which fueled the engine of colonial modernization.[25]

Drawing on Michel Foucault's theory of biopolitics, scholars such as Nikolas Rose challenge the distinction between the personal and political spheres by maintaining that the creation and evolution of personal lives and ideals of individualism were not unaffected by the interventions of nation states. Although the realms of intimate lives, desires, and subjective experiences were perhaps the only places in which to locate private selves, Rose asserts that these arenas were intensively governed. The rise of the social sciences in the early twentieth century, or what Rose calls "psy" in a variety of human technologies, allowed the public to have greater access to the personal. With the aid of this knowledge, states increasingly engendered and governed "the self" through institutions such as schools and factories.[26] Whereas in some traditions the concept of society has been characteristically linked to "constraint," the structural properties of modern institutions have been found to be both constraining and enabling. Neither individual nor structural agency was absolute but was characterized by degrees of accommodation. Eschewing the confines of the subject-object dichotomy or the "double hermeneutic," authors such as Anthony Giddens propose that individuals in modern societies have been and remain "social theorists," whose theories helped constitute structural activities and institutions.[27] Thus, the transformation of Korean culture did not necessarily mean that the ideals and practices of early twentieth-century Koreans were implanted by Japanese settlers. Regardless of their antagonistic political goals, Korean nationalists and Japanese colonists "with their mutual endorsement of capitalist modernity shared much in their historical understanding and approaches to Korean national culture."[28]

Japan's diplomatic opening of Korea in 1876 paved the way for not only colonial governmentality but also the expansion of capitalism, and with it, the exploitation and rationalization of its natural and human resources. By 1882, only six years after the signing of Japan's first commercial treaty with Korea, the total value of trade between the two countries had increased by twenty-fold as compared to the previous decade. Open entrepreneurship on the peninsula stimulated investment in modern economic institutions. In 1878, the Dai Ichi Bank established a branch in Korea; by 1905, there were three Japanese- and two Korean-owned banks. Similarly, railroad construction was initiated in the late nineteenth century and, by 1906, the Ŭiju-Seoul-Pusan line, vertically connecting the peninsula, was complete.[29] The overall standardization of currency (wŏn; yen) was accomplished a year after annexation in 1911. Although the focus of Korean production was on agriculture in the first decade of colonial rule, the WWI economic boom fueled the founding of modern mills. The number of factories on the peninsula grew exponentially from approximately 250 in 1911 to around 4,000 by 1927.[30] "Rationalization," as expressed in the technologies and organization of Koreans, corresponded with the rise of industrial capitalism in early twentieth-century Korea.

Traditional economists perceived the transition from the household to the light industries and then to the heavy industries as a procedure of replacement, where the latter displaces the former.[31] The rise of heavy industries was equated with the augmentation of the male workforce and the primacy of masculinity in the labor market. These scholars, by underscoring female employment in the first phases of modern economic development, also depict women as interlopers who did not stay in the wage-earning labor force. Although nations modernizing earlier might have experienced distinct stages of industrialization, in countries that mechanized later, these stages overlapped. With rapid industrialization, heavy industries and large-scale factories did not deter but rather spurred the maturation of light, small-scale businesses, which frequently sold roughly processed goods to larger manufacturing plants that completed the production. Because factory women have been largely circumscribed within light industrial labor during early development, the implications of faster economic growth for female labor are profound. If accelerated development blurred the stages of mechanization, male workers in heavy industries might not have so conspicuously supplanted women in light industries.

The effects of historical timing, as well as colonial investment and the establishment of a "techno-administrative state," in shaping women's work can be gauged in part through an investigation of early twentieth-century mechanization and the rise of women's factory labor in colonial Korea.[32] Modernization was a progression that fluctuated considerably throughout the thirty-five years of colonial rule. As a result, employment opportunities for women evolved with changing cultural, economic, and political surroundings. Textile and silk mills offered women jobs throughout the 1920s and early 1930s but, during this time, women also commenced other forms of industrial and agricultural wage work. As urban centers grew, women began work in rubber mills, ceramic workshops, and the flourishing service sector. With rising female participation in public education and vocational training, a select number of Korean women rose to white-collar positions, and by the end of the colonial era, women came to be represented in virtually every profession. Whereas in the first half of the colonial era, women's work was associated with unskilled labor in household and light industries, by the Second Sino-Japanese War, women's work often meant trained labor in heavy industries, as well as skilled positions. The expansion of public education also prepared youth for white-collar positions as clerks, nurses, salespersons, teachers, telephone operators, and typists.

## Political and Economic Changes in Early Twentieth-Century Korea

From the point of Korean-Japanese annexation (Han-Il happyŏng) in 1910, the preliminary efforts of the Government-General were centered on establishing imperial authority. To assess the status of commerce, natural resources, and population, the collection of year-end statistical records on agriculture, industry, and demography began in 1911. In the same year, an act was passed in the Japanese National Assembly, or the Diet, which allowed the governor-general to issue ordinances with the same effect as Diet acts. In 1912, the governor-general issued edicts on civil and penal matters in Korea, instating in the peninsula major Japanese laws, such as the Civil Code, Commercial Code, Criminal Code, and the Codes of Civil and Criminal Procedures. Initial orders installed civil guidelines, but they were also economically motivated, for these adoptions paved pathways for the private, often Japanese, ownership of Korean land. The cadastral land survey (1910–18) assessed the peninsula with reference to the ownership, value, and quality of land to provide information for investors and to procure "government revenue through taxation."[33]

Japanese endeavors for capital accumulation in Korea, through the formation of banks, insurance companies, and other financial institutions, began in the late 1870s with events such as the founding of the Daiichi Bank in Pusan. By 1905, the old Korean currency was retracted and new coins that corresponded with the Japanese yen were introduced. Although fiscal reforms implied compatibility for Japanese and Korean investors alike, the Company Law of 1910, which required all new businesses to obtain official permission, placed Korean entrepreneurs at a disadvantage. Alongside fiscal reforms, infrastructural projects such as the building of roads, port facilities, and communication networks commenced after 1910. Nonetheless, the colonial government initially espoused a policy of deliberate underdevelopment to "preserve Korea as a secure market for Japanese manufactured goods."[34]

As Japan's economy transitioned from having a net surplus of food grains to importing its rice in the late nineteenth century, the Meiji government planned to maximize agricultural production in Korea. Cultivation policies during the first decade of colonial rule were aimed at gradual rather than expeditious growth. These moderate plans, however, were reconsidered as skyrocketing inflation caused discontent among rural and urban workers in Japan, prompting the rice riots of 1918. A thirty-year plan for increasing Korean rice production was introduced, and rice prices in Japan and Korea were immediately halved. Although the sharp fall of rice prices led to disturbances in the countryside, the economic opportunity prompted by the First World War shifted Japan's position from a debtor to a creditor country. The abolition of the restrictive Company Law in 1920 and the formation of "cultural rule" (bunka seiji) promoted Korean investment in nonagricultural sectors.[35]

The favorable economic conditions sparked by WWI demand served as a catalyst for capital accumulation and industrial development in Korea. As Nakamura Takafusa described, "the wave of industrialization and incorporation brought by World War I was also the beginning of industrialization in the hitherto suppressed colonies."[36] Between 1917 and 1919, large corporations with over one million yen in invested capital were founded for the first time in Korea, among them the Chōsen Textile Company in Pusan, Mitsubishi Ironworks in Kyŏmip'o in Hwanghae Province, and the Onoda Cement Factory in Sŭnghori in South P'yŏngan Province, and in Seoul the Chosŏn Silk Mill and the Kyŏngsŏng Textile Corporation.[37] The 1920s witnessed the burgeoning of manufacturing, as the number of factories rose from 252 in 1911, to 2,384 in 1921, to 4,261 by 1930.[38] Technologically intensive projects were also instituted in Korea, including the Japanese Nitrogen Fertilizer Company (Nihōn chissō hiryō kabushiki kaisha) and the Korean Hydroelectric Company (Chōsen suiryoku tenki kabushiki kaisha). Japanese corporate giants built bases in Korea throughout the 1920s, but at the end of the decade, factories with over 100 workers comprised only 2 percent of the total, medium-scale factories 3 percent, and the remaining 95 percent of the factories were small-scale enterprises, often employing fewer than five workers.[39] Despite a few large companies, most colonial businesses stayed small.

As Sang-Chul Suh determines, modern economic development in colonial Korea took four distinct stages: infrastructural and bureaucratic preparation before 1919; the advancement of commercial agriculture and light industries during the 1920s; the founding of heavy industries as well as the further expansion of light industries during the 1930s; and war mobilization between 1937 and 1945. Drawing on such models, many economic historians have emphasized the explosive growth of heavy industries, such as metallurgy, machine and machine tools, chemicals, ceramics, as well as gas and electricity in the late 1930s.[40] They have implied that while small- to medium-scale enterprises prospered from the late 1910s to the 1930s, by the late 1930s, manufacturing franchises in Korea prioritized the production of capital goods over consumer goods. The fact that heavy industries accounted for a greater proportion of total output, however, did not mean that light industrial production decreased. Although participation in household industries and small-scale enterprises, such as rice polishing and food processing, declined gradually throughout the 1930s, cotton textile, silk, and rubber manufacturing projects, evermore reliant on female labor, continued to grow throughout the colonial era.

## The Transformation of the Korean Workforce

Although scholars have explored the extent of production in colonial Korea, detailed population studies have been hindered by inconsistencies in source material. As noted by Andrew Grajdanzev, the discrepancies between the year-end estimates of the Government-General (Chōsen sōtokufu tōkei nenpō) and official census records (Chōsen kokusei chōsa hōkoku) cause much confusion over the exact population figures of colonial Korea. Of the two sets of data, the census records, begun in 1925 and taken every five years, are known to be more accurate.[41] Viewed in conjunction with statistical records, these indicate that Korea's population grew from approximately fifteen million in 1910 to twenty-five million by 1944, with an average growth rate of 1.8 to 1.9 percent a year.[42]

Population growth coincided with urban migration. Over 85 percent of this burgeoning Korean populace was engaged in farming in the early 1920s. The proportions of agricultural workers decreased, however, to 78.4 percent in 1930 and to 72.7 percent by 1940. While the urban labor force was provided in part by natural population growth, the influx of wage workers from the impoverished countryside also contributed to the nonagricultural workforce. As Gi-Wook Shin and others have explained, the introduction of commercialized farming, prompted by "high prices, growing exports and expanding markets," drastically altered the composition, distribution, and operations of the rural labor force.[43]

For most Koreans in the early twentieth century, staying in the countryside was anything but secure.[44] Although the majority of Korean landlords were small-scale cultivators and village residents, commercialized agricultural production emerged in the 1910s and 1920s as large Japanese landholders invested in technological resources and hired farm agents and managers. Thus, tenancy increased on the whole throughout the 1920s and 1930s, and more rural families came to depend on outside wage work. Approximately two-thirds of the rural population rented part or all of its land, so landless tenants and even semi-tenants worked for wages in commercial farms. By 1930, seasonal workers composed 48.3 percent of the entire agricultural workforce. Throughout the colonial era, the gap between small-scale, "poor farmers" (*pinnong*) and "prosperous farmers" (*punong*) widened.[45]

Although peasants could have supplemented their incomes through agricultural wage work or semi-industrial labor in rice-polishing mills, breweries, or handicraft workshops, as factories proliferated throughout the 1920s and 1930s, urban migration became an increasingly viable alternative. In 1910, only eleven cities had populations of "14,000 or more and the aggregate urban population was only 566,000," or about 4 percent of the total Korean population. But by 1940, an approximate 2.8 million urban dwellers formed 10.5 percent of the total peninsular population. By 1938, fifty cities had 15,000 or more residents, with the most pronounced rates of growth in Seoul (Kyŏngsŏng or Keijō), Pusan (Fusan), and Pyongyang (Heijō). The population of Seoul, 343,000 in 1925, more than tripled to 1.1 million by 1940. As Hori Kazuo asserts, Seoul in 1940, housing 19.7 percent of Korea's total urban population, was comparable to contemporary Tokyo, containing 22.5 percent of Japan's total urban population.[46] As northern metropolitan centers such as Pyongyang and Wŏnsan flourished, disenfranchised farmers, concentrated in the south, migrated north in pursuit of steady employment. The disparities in the rates of population growth between the two regions

*Plate 1.1*   An open-air cotton storage in Mokp'o (1920s).

*Plate 1.2*   Cotton market in Seoul (1930s).

*Plate 1.3*    Downtown Seoul (1920s).

*Plate 1.4*    Downtown Wŏnsan (1930s).

*Plate 1.5*    The Ch'ungmuro (Honmachi) shopping district in Seoul (1930s).

were marked; between the 1925 and 1940 censuses, the urban population in the southern provinces rose by 298 percent whereas cities in the northern provinces grew by 445 percent.[47]

Apart from urban and interregional migration within the peninsula, a unique feature of colonial Korean industrialization was overseas migration, particularly to Manchuria and Japan. Throughout the late Chosŏn dynasty and in the early years of colonial rule, and frequently after bad harvests, farmers from the northern provinces crossed the Yalu and Tumen rivers into Manchuria and the Russian Far East in search of fertile land. Areas of settlement were the highlands of Manchuria, the north Kando area, and

Siberia. Migration to Manchuria, nonetheless, decreased by the late 1920s as the Chinese started to view "Korean migrants as civilian front units carrying out the economic penetration policy of the Japanese Kwantung Army." After the Manchurian Incident in 1931 and the founding of Manchukuo in 1932, however, Korean migration surged. Whereas approximately 460,000 Koreans resided in Manchuria in 1920, this figure rose to 607,000 by 1930 and to about two million by 1945.[48]

As Koreans of the northern provinces crossed the borders to Manchuria during the early years of colonization, rural people of the southern provinces began migrating to Japan in the late 1910s. Korean immigration to the islands was an issue of debate as Japanese unemployment soared in the early 1920s. But, by the late 1920s, the number of Koreans in Japan increased even as migrants came under heightened government control. By 1945, an estimated 2.4 million Koreans resided in Japan. From the late nineteenth to the early twentieth centuries, Koreans also settled in Russia and Hawaii. Between 1903 and 1905, approximately 7,000 Koreans migrated to work in Hawaiian sugar plantations, although emigration halted with Japanese intervention in 1905. While in the early twentieth century there were 20,000 Koreans in the Russian Maritime Province, by 1945, this population multiplied to 150,000.[49] Despite high rates of overseas migration, the vast majority of workers remained in Korea. As cities, institutions, and forms of employment multiplied throughout the colonial era, more and more rural peasants looked toward factories in search of work. Industrialization, commercialization, urbanization, and the displacement of the agricultural workforce fundamentally altered the structures and patterns of the Korean economy. These procedures of destruction and reconstruction were crucial for the formation of an industrial labor force.

*Female Factory Labor in Colonial Korea*

The first generation of wage-earning women in colonial Korea encountered a wide array of work settings. Like factory operatives elsewhere, Korean women commenced wage work in small- to medium-scale projects such as spinning and weaving, silk, rice polishing, and food-processing mills, enduring twelve- to fourteen-hour workdays and around-the-clock supervision.

During the first half of the colonial era, businesses usually took the form of small-scale mills manufacturing consumer goods such as knitted socks, tobacco, and matches. Small-scale factories often employed day laborers seasonally, as did food-processing and rubber shoemaking plants. The employees of these enterprises sometimes engaged in agricultural wage work in the spring and summer months and in factory labor in the autumn and winter months. The conditions of mill work in cities such as Pyongyang, especially throughout much of the 1920s, resembled those of contemporary Tianjin, where household and large-scale businesses both flourished. In early twentieth-century Korea, as in North China, the home manufacture of "matchboxes and knit goods survived alongside the factories and indeed helped them prosper."[50]

Despite the speed of development, Korea's industrial climate throughout the first half of the twentieth century was varied, composed of handicraft industries, small workshops, and large factories. Cross-referencing records on workers in cotton textile and silk production, as well as those in heavy industries such as cement manufacturing, shows that the terms and conditions of employment in large-scale enterprises were dramatically dissimilar to those in small-scale industries. In small mills with fewer than 100 employees, workers were often just placed into skilled and unskilled groups. The division of labor was more intricate in large-scale plants that hired thousands. Japanese conglomerates with more than forty years of industrial experience instated in their factories in Korea complex systems of labor stratification. Entry-level employees of corporations such as the Kanegafuchi and Tōyō spinning and weaving companies, as well as of Onoda Cement, were generally recruited and housed in factory dormitories.[51] Therefore, the standards of factory work during the colonial era were far from uniform. Manufacturing in Korea did not evolve in a neat historical progression from household to workshop to modern mill. Instead, all three types of production emerged simultaneously, "with each of them sometimes helping to spur the other's growth."[52]

While total production in Korea lagged in the few years following the Depression, the Japanese occupation of Manchuria, completed in 1932, boosted growth. Most industries recovered by 1934, and heavy industries built plants at a faster pace in Korea and Manchuria.[53] Like their Japanese counterparts, Korean businesses also profited from Japan's venture into

Manchuria. An P'yŏngjik, Carter Eckert, and Soon-Won Park, among others, contend that by the late 1930s Koreans increasingly rose to skilled positions, providing new insights on the changing status of colonial workers. Although in many studies of colonial factory women little differentiation has been made between the workers of the late 1930s and those of earlier years, a critical evaluation of the wartime workforce suggests that both labor forces and markets experienced significant transformations during the colonial era.

State attention was first given to agriculture and light industries, but the mid-1930s ushered in heightened bureaucratic and corporate attention to the output of heavy industrial, capital goods. The Korean women who entered the factories from 1917 to 1937 most likely encountered rudimentary settings, but many of the succeeding generation experienced more sophisticated labor conditions and options previously unavailable to women. Accelerated industrialization throughout the 1930s allowed for some fortunate lower-class farm girls to attain skilled positions for the first time. After the 1931 enactment of the Major Industries Control Law in Japan, large corporations rapidly began building subsidiaries in Korea.[54] These colonial impulses and, just as important, the responses of local populations, dramatically altered the rate and character of Korean modernization as well as employment opportunities and labor conditions by the 1930s.

The female industrial labor market experienced undeniable expansion throughout the thirty-five years of Japanese rule. During the early phases of mechanization, employers of cotton textile projects usually assigned women to the less strenuous divisions in spinning and weaving factories, but they soon took over the entire sector. Jobs considered too demanding for women in the 1920s were given to them by the 1930s, as the Depression prompted companies to cut expenses. Because the hiring of women and children was the least costly option for owners, female employment increased across almost all sectors of production. In enterprises that already depended heavily on female labor, such as those manufacturing textiles and rubber shoes, women comprised up to 90 percent of those employed. In the 1920s, men dominated the textile divisions requiring strength, such as mixing and scutching, where large bales of raw cotton had to be carried. By the 1930s, however, women took over these tasks. By the start of the Second Sino-Japanese War, the percentage of women workers in textiles remained steady at over 80 percent. Nevertheless, proportions of women more than doubled in metalwork, tripled in gas and electricity, and increased by fourfold in the machine and machine tools sector.[55] By the last years of the colonial era, women and children were employed in all industries, manufacturing shoes, machine parts, airplane components, and munitions. Public schools also introduced to youth the possibility of obtaining white-collar jobs as clerks, nurses, operators, salespersons, stenographers, and teachers.

The employment of women in modern industries was fueled by public education, formally instituted with the 1911 Educational Ordinance of Chōsen. Nonetheless, pedagogical endeavors in the first decade of colonial rule were intended not to provide public education for all but to centralize schools. Regulations concerning private schools were amended in 1915, resulting in the closure of many private schools. In 1910, missionaries operated 1,973 private schools, whereas by March 1, 1919, there were 690 such institutions left.[56] The number of traditional Confucian academies for young men similarly declined, showing the efficacy of the Government-General in eliminating private education, and pedagogical competition, in Korea.

Enrollment in state schools did not become widespread until 1922, after the institution of the Revised Educational Ordinance that reformed the public education systems of Taiwan and Korea. Ordinary schools (*pot'ong hakkyo*; *futsu gakkō*) modeled after Japanese and Taiwanese schools were built, offering students four years of instruction.[57] Throughout the 1930s, popular involvement in education grew exponentially. While universal education improved options for some, its impact varied. Many children were unable to attend schools due to material circumstances and, even if enrolled, could not always stay. Because children did not necessarily start schooling at a uniform age, classes frequently included students of different ages.

Although some women's historians in Korea have inflated the impact of mission education on the women's movement and female activism, a comprehensive investigation of early twentieth-century education conveys that those trained in Protestant institutions comprised only a small minority. By the 1920s, the predominance of mission schools diminished, and students received instruction in colonial institutions. Although less than 1 percent of school-aged girls were enrolled in public schools in 1918, by 1933, roughly

*Plate 1.6*    Weaving class in a school for girls (1920s).

20 percent of all elementary school–aged girls were registered in more than two thousand ordinary schools. By 1939, primary schools enrolled 306,000 girls, accounting for 25 percent of the female, school-aged population in Korea.[58] The public education system became more comprehensive throughout the 1930s and by the latter years of the colonial era, schools became the training and recruiting grounds of future employees.

## Summation

According to upholders of the "big bang" theory of industrialization, modern economic development was universal and predictable. Similarly, contemporaries and intellectuals rendered and reproduced portraits of "factory girls" that made their historical, cultural, and contextual diversities indistinguishable. The cotton textile industry was seen as a bridge between the two modes of production: household and industrial.[59] Following these interpretations, some historians of colonial Korea have argued that although women employed in light industries, such as spinning and weaving, silk reeling, and rice and food processing, were prominent in the early stages of mechanization during the 1920s, women's productive influence depreciated as entrepreneurial emphasis shifted to the heavy industries in the 1930s.[60] Such views assume that modernization was a linear progression, with large-scale factories displacing all older forms of manufacturing. Nonetheless, methods of industrial production were mediated by the paces of economic development. Women's entrance into factory labor was not an inevitable outcome of mechanization. Textiles and other light industries were predominant in Great Britain for several decades, but in later-industrializing United States and Japan, manufacturing priorities shifted more quickly from the production of consumer to capital goods. In countries mechanizing even later still, this transition from light to heavy industrial production was not only faster but more heterogeneous. In Korea, where industrial strategy was largely affected by experienced Japanese corporations and government promotion, the rate of economic development surpassed the pace of earlier modernizers.

Reviewing the consistencies and contingencies of modernization and exploring the variations of women's wage work across times and cultures underscores the importance of economic and cultural specificity in determining the characteristics of women's industrial labor. Although still diverse, cross-cultural analysis indicates that Korean women workers, particularly those employed in large-scale factories, confronted rigorous standards of recruitment, training, and labor control. Therefore, insofar as they lived outside of factory compounds, working women in European and Chinese cities might have been "better off." But the testimonies of and records on Korean factory women demonstrate that residences were also places of socialization, collective resistance, and personal growth. If accommodation implies servility, Korean workers were more submissive than their historical counterparts. If by their accommodation Koreans secured long-term opportunities, however, their adaptability was a powerful survival technique. Adaptation might have been fruitful in the long run because, unlike the pace of development in other regions, industrialization in Korea advanced ever more rapidly in the last fifteen years of colonial rule.

## Modernization and the Rise of Women's Wage Work

At the turn of the twentieth century, suffragists across cultures depicted the rise of female industrial labor in novel terms, but as Ada Heather Biggs observed in 1894, the share taken by women in the "work of the world" had not altered in amount or in intensity, "only in character."[1] In colonial Korea, modernization offered new forms of work but did not drastically transform the extent of women's obligations. Although capitalist development was associated with the decline of household manufacturing, progress was uneven and homes, workshops, and factories existed in harmony for much of the early twentieth century. As they did in the Chosŏn dynasty, most women in colonial Korea engaged in domestic chores, and married women often served as managers of handicraft production. By the 1920s, however, women began to venture into agricultural wage work on commercialized farms and paid labor in factories and workshops producing consumer goods. While unmarried women remained in domestic service and mill work, increasing enrollment in public education opened to female applicants skilled positions such as clerks, teachers, and typists by the latter years of the colonial era. When examined in conjunction, commercialization, industrialization, and the burgeoning of women's wage work indicate that modern factories, though important, offered only limited forms of employment among a variety of women's occupations in early twentieth-century Korea.[2]

According to Friedrich Engels and Marxist feminists, the genesis of the family, the state, private property, and gendered divisions of labor corresponded with the accession of the agricultural mode of production. Thus, most traditional societies organized around agrarian households, including Korea at the turn of the twentieth century, produced goods domestically in a system that Joan Scott and Louise Tilly call the "family economy." Because specialization made production more efficient, a married couple maximized their yield from sharp divisions of labor because the husband could "specialize in some types of human capital and the wife in others."[3] The successful reproduction and rearing of children offered parents additional labor for farming and home manufacturing projects. Although the size and composition of the household relied on demographic, economic, and social conditions, the specific requirements of individual households were also significant since they "put practical constraints on the permissible variation in household age and sex composition." Therefore, parents raised the number of children they needed and rarely more. Parents also synchronized parturition to optimize their old age security. In early twentieth-century Korea, where primogeniture was common, firstborn sons, and not daughters, continued the patriline. While peasant families might not have had much to leave to descendants of either sex, girls, who were legally bound to the families of their future husbands, seldom received inheritance and were often seen as deficits, not assets, to the family economy. The productive demands of the household determined not only family dynamics and structure but also the frequency and timing of familial events. Births, periods of education or training, employment, marriage, moving, and retirement were all mediated by the changing needs of the household.[4]

This chapter extends Biggs's observation to the colonial Korean context: although female contributions to political economies seemed unprecedented, women's work in family economies had persisted since prehistory. The transition from the agrarian to the mechanized mode of production did not engender women's work but merely differentiated between paid and unpaid forms of labor, arguably diminishing the value of domestic labor altogether. To portray how modern economic development affected women in early twentieth-century Korea, I first examine the family economy, or the domestic mode of production organized around households. A woman's role in household production depended on her marital status and life stage; responsibilities also differed for mothers and daughters. While the family remained a productive unit, the structural transformations of labor brought on by industrialization and commercialization necessitated the familial adoption of the wage system. I then review how these economic progressions reconfigured the living and working conditions of married and unmarried women. Although prominent features of modernization, including the expansion of monetary exchange and paid labor, blurred the distinctions between the duties of older and younger women, wages did not alleviate female obligations in family economies. Apart from performing unpaid domestic tasks, married women took on jobs flexible enough for simultaneous child care, such as household handicrafts, agricultural wage work, urban service work, and labor in small- to medium-scale food-processing and rubber factories that hired part-time and seasonal workers. Unmarried women, though continuing their supportive roles in family economies, experienced new forms of recruitment and training. Some young women remained in domestic service whereas others were mobilized for work through state employment offices. Nevertheless, the majority of poor, single women entered the factories, as light industrialists favored the employment of women in their teens and early twenties. This development coincides with the last issues of inquiry: the emergence and expansion of female labor-intensive enterprises in the 1920s and 1930s. Finally, I attempt to interpret why these women entered the factories—their motivations and rationales—all of which strongly suggest that women's contributions to political economies were driven by familial, personal ties.

*The Family Economy in Colonial Korea*

At the outset of Japanese annexation in 1910, the most pervasive form of capital in Korea was land and its cultivated products. Landlords, owner-cultivators, semi-tenants, landless tenants, and wage workers constituted rural society. Although Korean farmers lost their property in ever greater numbers throughout the nineteenth century, based on the cadastral land survey of 1910–18 that recorded, titled, and taxed each acre of land, over 77 percent of the households in Korea lived on their land as tenants or semi-tenants. Landless, or full, tenancy rose throughout the colonial era; in 1922, tenants comprised 41 percent of the rural workforce, and this figure climbed to more than 60 percent by the early 1940s. In the countryside, tenant farming continued but did not yield enough for many families to meet their material needs. Whereas in 1913, tenants who also worked for wages comprised 13 percent of the rural workforce, "by 1926, the figure was 16 percent," and by 1940, over 35 percent.[5] Therefore, members of tenant households not only cultivated borrowed land but also worked as migrant wage farmers or as operatives in mills, as both forms of employment increased during the colonial era.

Whereas in premodern economies, men's and women's work was defined by the labor requirements of the household, with the ascent of the monetary system, the functions of individual family members came to be shaped by the family's need for income to offset living expenses. In addition, with industrial and urban development, the location of work changed. But the modern "family wage economy" served as an extension of the family economy "as the old rules continued to operate in new contexts." Wages neither initially nor totally replaced the household mode of production, as family strategies were not directly proportional to land cultivation. A household with a small amount of land, for example, may have needed "large amounts of labor because the return to labor [was] lower on tenancies than on land owned by one's own household."[6] Although married and unmarried men continued to farm as tenants, many also became hired workers. According to Gi-Wook Shin, more than 22,600 tenants hired more than 81,800 wage laborers in 1930. As commercialized rice production prompted agricultural restructuring, more men and women farmed for wages throughout the 1920s and 1930s. Those women in the countryside without paid employment continued spinning, weaving, and knitting, as well as pickling vegetables and making straw bags and other handicrafts to supplement household earnings. In assessing whether "work improved women's positions," Scott and Tilly conclude that although mechanization reformed the characteristics and conditions of labor, women's economic duties did not alter as dramatically as previously thought.[7]

While people's economic, or functional, responsibilities remained the same, how they fulfilled these commitments transformed throughout early twentieth-century Korea. Whether urban or rural, the proliferation of wage work differentiated individual and family schedules, rationalizing and dichotomizing labor and leisure time. The commercialization of agriculture and industrial development fueled urban migration as the number of the peninsula's city-dwellers more than quadrupled between 1910 and 1945. Over 60,000 Koreans left their farms between 1930 and 1935; in the latter half of the 1930s, over 220,000 moved out of the countryside. Most migrants found jobs on the peninsula, but throughout the 1930s and the Second Sino-Japanese War, millions of Koreans went abroad to Japan, Manchuria, and Siberia. Technology and circumstance determined not only individual but household behavior whose goal was always adaptation, or "the solution of problems caused by the environment."[8]

Although some families moved as a unit, more often family members left households individually for seasonal or contractual wage work. According to Carol Stack and Linda Burton, the movement of members within a household can be defined by: "kin-work," or the labor that families need for their continuation; "kin-time," or the temporal and sequential ordering of family transitions; and "kin-scription," or "the process of assigning kin-work to family members."[9] As the backbones of still largely agricultural communities, parents and grandparents usually stayed behind in the countryside, engaging in farming and household manufacturing. The few children remaining at home or sent away for schooling were most likely primary sons. While industrial labor was unprecedented, children's contribution to the family economy was not. The practice of hiring out second- and third-born sons as field hands and daughters as domestic servants was steeped in tradition. Nevertheless, there were noticeable discrepancies in how parents scripted, or appropriated, their offspring for production. A detailed analysis of urban migration in the 1920s and 1930s shows that more women, in particular young women, moved to cities as compared to men. From 1925 to 1930, children aged five to nine comprised 21 percent of the female and 19 percent of the male populations migrating to cities; 35 percent of all females and 29 percent of all males moving to urban centers were aged ten to fourteen. Between 1930 and 1935, children aged five to nine composed 18 percent of the female and 15 percent of the male migrants to cities; and 31 percent of all females and 24 percent of all males moving to cities were aged ten to fourteen. Although more men migrated after the age of twenty, more women, on the whole, made the rural-urban pilgrimage. Between 1925 and 1930, over 22 percent of the female and 19 percent of the male populations moved from the countryside to the city. Census records also show that between 1925 and 1935, women and minors under fifteen years of age entered urban factory employment at faster rates than adult men.[10]

The demographic shifts seen throughout the colonial era were both general and local. In 1930, the rate of childbirth among 1,000 women ranged from 41.7 to 42.7; by 1935, this figure decreased from 42.7 to 39, and in 1940, from 39 to 35.7. The age of marriage among women also rose slightly, from 17 years of age in 1930 to 17.5 in 1940. Patterns of marriage and childbirth also varied according to an individual's age and his or her place of residence. In 1930, 31 percent of rural women between the ages of fifteen to nineteen were without mates, whereas 64 percent of urban women in the same age range were single. Whereas less than 1 percent of rural women aged twenty to twenty-four were single, roughly 16 percent of urban women in the same age range remained unmarried. Modernization—synchronizing the procedures of capitalization, colonization, industrialization, rationalization, and urbanization—transformed the settings of life and work and altered the characteristics of human alliances, making them more fluid and multifaceted than premodern ties. Although some women were more affected by these changes than others, for most of the colonial era, females constituted over 30 percent of the industrial workforce and almost 40 percent of the total workforce, demonstrating that modern economic development owed much to the labors of women.[11]

## Married Women's Wage Work

The work of a married woman (*kihon yŏsŏng*) in premodern Korea depended on her husband's economic position, his landholdings, and his status. Aristocratic women were indoctrinated with strict codes of behavior, but the extent to which women of the lower classes aspired toward elite values is difficult to measure. In most rural households, a vast array of tasks ensured the survival of their inhabitants: harvests had to be "processed and stored, domestic animals cared for, meals made and served, clothing provided and maintained." As Clark Sorensen explains, "a clear division of labor for males and females" prevailed for the accomplishment of these tasks. Men's work consisted of outdoor labor: construction, farming major crops, operating irrigation systems, and maintaining farm tools as well as organizing markets exchanging goods and labor. Married women usually performed the "inside work" of the home: "producing and managing food stores, growing vegetables, preparing meals, making clothing and bearing and raising children."[12] Foodstuffs, crafts, and apparel manufactured by women at home were bartered or sold in markets, reflecting their multiple functions. Weaving was done in almost every household, and portions of traditional state taxes were levied in cloth. Thus, by spinning and weaving, preparing and exchanging vegetable goods, and raising farm animals, women held pivotal roles in sustaining premodern economies.

Although rural women continued their household chores, ever more married women ventured into agricultural wage work throughout the colonial era. In premodern society, tilling the fields (*nonggyŏng*) or outdoor labor (*ogwoe nodong*) was reserved for men. Nonetheless, Government-General policies geared toward the mobilization and stratification of rural labor, including the Rural Revitalization Campaign (Nongch'on chinhŭng undong) of 1932, reconfigured the gender boundaries of farmwork. By the 1930s, through membership in agricultural cooperatives (*kongdong kyŏngjak*), women commenced paid, outdoor labor in large numbers. According to the 1930 census, females not only comprised around a third of the industrial workforce but also formed 32.2 percent of the aggregate number of tillage farmers and over a third of the rural labor force.[13]

Women's employment on farms took a variety of forms, but as in manufacturing, married or older women were represented in certain types of agriculture more than in others. Whereas females comprised only 2.9 percent of the labor force in forestry (*imŏp*), they were predominant in animal husbandry (*ch'uksan* or *mokch'uk*), forming 76.4 percent of the workforce. Women raised chicken, pigs, sheep, rabbits, and other domestic animals (*kach'uk*) for meat, eggs, leather, and wool.[14] Sericulture was done almost exclusively by women, who constituted 95.2 percent of the total workforce in silkworm farming (*yangjam nongga*). In vegetable cultivation (*singmul chaebae*), women grew bean sprouts, bellflowers, bracken, spinach, turnips, and other produce. Although most of the raw cotton (*yugjimyŏn*) processed in colonial spinning and weaving mills was imported, the cotton harvested (*myŏnjak*) in Korea was handled by women. Also in the 1930s, the cultivation of medicinal plants (*yagyong singmul*) was launched as a female labor-intensive project.[15]

Were rural women "better off" than city women? The conflicting data makes this question difficult to answer. As in the factories, laborers on commercial farms were subject to a wage system that systematically reduced salaries according to their ethnicity and sex. Throughout the colonial era, Japanese workers' wages were double that of Koreans, and men earned twice as much as women. In general, agricultural earnings were lower than industrial ones. For example, in 1923, the average daily wage in mining and construction was 0.88 and 0.94 yen, whereas in farming it was 0.45 yen. Moreover, earnings did not escalate with experience. Between 1925 and 1930, agricultural wages actually decreased from 0.81 to 0.73 yen for men and from 0.44 to 0.42 yen for women. A day's pay depended on not only the kind of labor performed but also the location of work. In suburban areas, farmers could earn as much as manufacturers. In the remote countryside, however, men in 1933 earned on average 0.35 yen, women 0.20 yen, and children 0.15 yen. Despite the low wages, agricultural work might have offered fewer restrictions and more freedom of movement than factory work, which makes qualitative comparisons difficult.[16]

Colonial modernization sought to turn rural craftswomen into agricultural and industrial laborers. Intended to mobilize women in the countryside, agrarian societies such as the Women's Labor Corps (Puin kŭllodan), the Women's Cooperative for Cotton Cultivation (Puin kongdong myŏnjakdan), and the Women's Youth Corps (Yŏja chŏngnyŏndan) burgeoned in the late 1920s and the 1930s. In addition, training centers (*kangsŭpso*) for specialized farming, such as sericulture and cotton cultivation, emerged and expanded.[17] During the early 1930s, nurseries for working mothers were built. By 1935, there were ninety-six rural nurseries (Nongch'on t'agaso) in Kyŏnggi Province, 100 in South Kyŏngsang Province, and 149 nurseries in South Chŏlla Province.[18] But the establishment of nurseries did not necessarily translate into their adequacy, as rural women were frequently found tilling the fields with "children on their backs and stomachs."[19]

Whereas in premodern Korea, the goods needed in daily life were produced in homes, with mechanization, commodities were manufactured in

*Plate 2.1*   Farm women selecting cotton (1930s).

modern mills. This transition was not immediate, and though large-scale projects multiplied during the colonial era, household enterprises remained widespread. In 1933, households were responsible for 40.1 percent of Korea's total industrial output. This figure decreased to 33.1 percent by 1935 and to 24.7 percent by 1938. Despite its decline, household production composed a formidable proportion of the peninsula's output throughout the colonial era. Officials recorded both men and women as the owners of household businesses, but the involvement of married women in such domestic projects was usually greater.[20]

According to the Government-General, scale defined a household industry as having five or fewer workers. Throughout the colonial era, a diverse array of goods, representative of nearly every sector of manufacturing, was also produced domestically. In food processing, women made alcoholic beverages and other provisions including bean paste, confectioneries, flour, soy sauce, and fish and vegetable preserves. In lumber, casks, furniture, and tubs were produced in handicraft workshops. Items assembled in the miscellaneous

*Plate 2.2*    Silk reeling in a household in colonial Korea.

category were "bamboo wear, paper ware, rattan ware, rush ware, needle work, footwear," and straw ware, such as ropes and bags. In textiles, household production involved the rearing of wild silkworm cocoons and silk reeling, as well as the spinning, weaving, and knitting of cotton and hemp. In ceramics, women molded bricks, earthen utensils, and tiles. In machinery, handicraft workers shaped farming tools and implements, and the chemical

*Plate 2.3* Cotton spinning in a household in colonial Korea.

material made in households included animal fat products, coal bricks, medicines, paper, fish fertilizers, as well as fish and vegetable oil.[21]

Domestic enterprises were regionally distinctive. Like commercial farms, there were more household industries in the southern than in the northern provinces. In 1933, approximately 6.5 million households (*ho*) were registered as producing handicrafts. By 1937 nearly 2,700 households, with more than 10,000 members, participated in household businesses in Seoul alone. Primary sources suggest that although rural households manufactured handicrafts as "secondary enterprises" (*puŏp*), or projects secondary to agricultural production, home businesses in cities produced goods as "main enterprises" (*chŏnŏp*), with handicrafts as the primary source of income. As access to vocational schools for the teaching of sericulture, knitting, and other skills improved by the mid-1930s, household industries in cities flourished.[22]

Rapid modernization in the 1920s offered married women jobs in some industrial sectors and settings, but in cities, there were other ways to make money. If women absconded or were dismissed from companies, they could remain in the city and work independently as peddlers or street vendors.

Rather than starting businesses on their own, however, more women probably looked to the service sector for employment. Lodges and motels hired women as cleaners and laundresses while restaurants paid them to prepare and cook food. For both the married and the unmarried, often sexualized, service jobs as waitresses or as servers in wine houses increased throughout the 1920s and 1930s, as women entered the rapidly expanding service sector in cities such as Inch'ŏn, Masan, Pusan, Pyongyang, Seoul, Taegu, and Wŏnsan. Between 1920 and 1945, female employees dominated the service industry, comprised approximately 30 percent of the labor force in mechanized manufacturing, and formed anywhere from 30 to 40 percent of the total labor forces in other productive sectors such as agriculture.[23]

*Unmarried Women's Work*

Throughout centuries of Korean history, a man's residence was secure from birth, whereas a woman moved upon marriage. A woman was not considered a permanent member of her parents' household, and at best, obtained a secondary position in her husband's household after she bore a son. In less privileged households, it was not uncommon for girls as young as eight or nine to leave their homes to take service in another family's household. The most pervasive form of work for unmarried women (*mihon yŏsŏng*) in the traditional family economy was household service. A domestic servant (*kasa sayongin*) was a household dependent who worked in return for food, board, and wages, usually paid directly to the servant's parents. Peasant boys tended to work outdoors in the fields as seasonal laborers, whereas girls were confined to indoor work year-round. Service was the customary means by which households exchanged labor supply to fulfill the demand for consumption.[24]

According to the 1930 census, approximately 120,000 Koreans worked as domestic servants; females comprised 76 percent and 97.7 percent of these women were live-in servants. Approximately 57 percent of these women were children under fifteen years of age, but servants were not always unmarried.[25] During times of economic hardship and surplus labor, married women up to the age of thirty frequently parted from their husbands to take service in another's household. Although, in traditional society, women worked in the residences of local landlords and elites, domestic employment became regulated by the 1930s. In Seoul and other major cities, employment offices served as intermediaries between those requiring and seeking domestic service.[26]

Systems of payment, however, were far from uniform, and whether women were given monthly or yearly wages varied according to employers. Wages for domestic service were "anywhere from 2.5 yen to 9 yen per month," but cases existed when after prepayment (*sŏn'gŭm*), young women worked without pay (*mugŭp*). In many instances servants were verbally as well as physically abused and were vulnerable to dismissal without pay at any given time. Still, women ran away far more often than they were formally dismissed. If servants earned what they needed for dowries or nuptial expenses (*honsu*), they could marry. But, if they escaped without pay, they most likely looked to serve in other households or pursued alternative forms of work. Contemporary intellectual Kim Wŏnju feared that servants who faced excessive hardship and mistreatment frequently fell prey to "unsavory" forms of work in "red-light districts [*yullakka*]."[27]

Although the predominance of female child workers in spinning, weaving, and silk-reeling enterprises can be accredited to Japanese industrial designs, it is equally arguable that the modern responsibilities of peasant girls were drawn from Korean customs. Just as it was not uncommon for young rural women to take service in other households in premodern Korea, daughters were expected to contribute to the family economy throughout the colonial era. Even if they did not supplement household earnings, many parents might have preferred for their daughters to live away from home, thereby easing their domestic burdens. Kim Yŏngsŏn, a former clerical worker at the Kyŏngsŏng Textile Company in Yŏngdŭngp'o, attested that parents did not necessarily send their daughters to the factory with the expectation that wages would be sent home. Rather, they did so to reduce the size and consumption levels of the family.[28] In Europe, daughters functioned as the "arms of the family economy." In East Asia, daughters were the "wings" of the household—family members who were expected to "fly away" in due time.

Extending Confucian paternalism, managers in textile industries devised recruitment methods and forms of residence drawing on the tradition of domestic service. Because manufacturing relied on access to domestic and imported raw materials, most factories in colonial Korea were located in cities and metropolitan centers near ports that legitimized employers' practices of recruiting and housing their workers. Moreover, the need to organize workers around machinery that often operated twenty-four hours a day further warranted rigorous systems of labor and residence. Seeking obedient employees, company recruiters searched the hinterlands for peasant girls from especially impoverished areas. Young women between ten and twenty years of age composed the majority of the workforce in spinning, weaving, and silk-reeling mills.[29]

The better-known cotton textile and silk-reeling factories in colonial Korea were large-scale enterprises owned by Japanese conglomerates with decades of corporate expertise. By the time they advanced into the Korean market, these businesses had established efficient methods of management and production that depended on the availability of an unmarried female labor force. Although heavy industries might have relied on a permanent labor force with dedicated lifelong employees, spinning, weaving, and silk-reeling projects required no such loyalty. The high turnover rates (*idongnyul*) accommodated employers who preferred to disregard workers' long-term security and enjoyed the cost-effectiveness of child labor. Owners and managers adopted forms of discipline that simulated the paternalistic aspects of domestic service: they provided wages and residence as well as classes on sewing, knitting, and other handicrafts for a period of three to five years. To compete, Korean employers adopted similar systems of production, recruitment, and labor control in their spinning, weaving, and silk-reeling mills.

## *The Emergence of Female Labor-Intensive Industries*

From their inception in the late 1910s and early 1920s, many of the first modern factories in Korea, particularly those specializing in silk-reeling, cotton-spinning, weaving, and knitting, almost exclusively hired women. Other businesses producing consumer goods, such as tobacco, wine, and food and rice processing mills, also favored female employees. While fewer married women worked in mechanized settings, they found jobs in chemical enterprises, manufacturing rubber shoes, as well as processing fish and vegetable oils. An evaluation of statistics on industrial labor throughout the colonial era shows the gradual movement of women away from food processing and textiles into almost all other manufacturing sectors. Although some scholars have claimed that decreased official emphasis on textile production in the late 1930s translated directly into the general reduction of women in the workforce, a critical analysis of the figures shows that women became more proportionally distributed across industrial sectors (Table 2.1).

From the late 1910s to the early 1930s, manufacturing on the Korean peninsula consisted of the production of cheap, coarse, daily commodities for the domestic and export markets. Because small-scale enterprises did not entail large capital outlay or technical machinery, entrepreneurs preferred them in view of the plentiful Korean labor supply. Nonetheless, large-scale food- and beverage-processing projects such as Asahi Brewing (Asahi jōzō), Dai Nihon Sugar Refining (Dai Nihon seitō), and Manchu Flour Milling (Manshū seifun) also commenced during this time. Other early firms prioritized their investments on the making of consumer and apparel goods, with women dominating the manufacturing sectors, including leather (*pibok*) and rubber shoe (*gomusin*) production.[30]

The modern textile industry in Korea began with the production of silk. Investment in silk reeling (*chesa*) was most prominent from the late 1910s to the early 1930s during the first half of the colonial era. In the early 1930s, cotton spinning (*myŏnbangjŏk*) gained greater priority, but by the latter half of the decade the focus of textile production once again shifted to the manufacture of synthetic fibers (*injogyŏn* or *ingyŏn*). The institution of large-scale silk mills in Korea was influenced by the advances of Japanese conglomerates. The first large-scale textile factory in Korea, the Chōsen Spinning and Weaving Company (Chōsen bōseki kabushiki kaisha), a subsidiary of the Mitsui group, was established in 1917. Filatures quickly followed and in 1919, Chosŏn Silk Reeling (Chōsen seishi) founded a factory in Taegu, and Kunze Silk Reeling (Kunze seishi) formed bases in Taejŏn in 1926 and Ch'ŏngju in 1929. Other companies affiliated with the Mitsui group invested heavily on the peninsula. Tōyō Silk Reeling, for example, established plants in Chinhae in 1930 and in Sariwŏn and Pyongyang in 1933, whereas the Kanegafuchi Spinning and Weaving Company formed a silk-reeling mill in Seoul in November 1925 and factories in Kwangju in 1930 and in Ch'ŏrwŏn in 1933. Other well-known businesses were Katakura Silk

TABLE 2.1

Distribution of Male and Female Workers Across Industries (1921–43)

| | 1921 | | 1930 | | 1935 | | 1940 | | 1943 | |
|---|---|---|---|---|---|---|---|---|---|---|
| | M | F | M | F | M | F | M | F | M | F |
| Textiles | 3.9 | 28.1 | 7.7 | 51.6 | 6.3 | 49.0 | 6.5 | 46.0 | 9.8 | 59.4 |
| Metals | 27.9 | 0.4 | 5.4 | 0.2 | 5.9 | 0.2 | 8.9 | 0.9 | 14.5 | 2.0 |
| Machinery | 5.9 | 0.1 | 4.7 | 0.0 | 6.1 | 0.2 | 16.0 | 1.3 | 16.9 | 1.6 |
| Ceramics | 14.0 | 3.5 | 8.9 | 1.3 | 7.7 | 1.5 | 7.3 | 2.5 | 12.0 | 4.6 |
| Chemicals | 2.7 | 1.2 | 15.7 | 21.3 | 25.1 | 26.7 | 31.1 | 24.1 | 18.5 | 14.6 |
| Lumber | 3.7 | 0.0 | 4.8 | 0.0 | 5.2 | 0.1 | 6.9 | 0.9 | 9.8 | 1.8 |
| Printing and Publications | 5.6 | 0.3 | 7.3 | 0.3 | 5.8 | 0.4 | 3.8 | 0.5 | 3.4 | 0.8 |
| Food Processing | 23.8 | 53.7 | 40.3 | 22.2 | 33.4 | 18.3 | 14.4 | 13.5 | 8.9 | 8.4 |
| Gas and Electricity | 1.4 | 0.0 | 1.3 | 0.0 | 1.1 | 0.0 | 0.0 | 0.0 | 2.4 | 0.3 |
| Miscellaneous | 11.0 | 12.1 | 3.7 | 2.7 | 3.5 | 3.6 | 5.2 | 10.3 | 3.8 | 6.4 |
| Total | 100 | 100 | 100 | 100 | 100 | 100 | 100 | 100 | 100 | 100 |

Source: Chōsen sōtokufu, Chōsen sōtokufu tōkei nenpō [Statistical yearbook of the Government-General of Korea] (Keijō: Chōsen sōtokufu, 1921, 1930, 1935, 1940, 1943).*

* Chōsen sōtokufu gakumukyoku shakaika, Kaisha oyobi kōjō ni okeru rōdōsha chōsa [Survey of labor conditions in factories and mines] (Keijō: Chōsen sōtokufu, 1923). See also Yi Chŏngok, "Ilcheha kongŏp nodong," 163; Pak Kyŏngsik, Ilbon cheguk chuŭi ŭi Chosŏn chibae [The management of Korea during Japanese imperialism] (Seoul: Ch'ŏnga sinsŏ, 1986), 481.

*Plate 2.4*    Cannery on the east coast (1920s).

Reeling and Yamajū Silk Reeling, which built bases in Seoul, Pyongyang, Taegu, Ch'ŏnju, and Hamhŭng throughout the 1930s.[31]

Few modes of production better symbolized the industrial revolution than the cotton textile industry. It was of paramount importance not only in its productive value and trade capacity but also in its establishment of modern facilities and operations. The first cotton mills in Korea, however, were not like the large-scale, technologically advanced factories in Japan of the same period but were small-scale workshops that hired an average of six to seven employees. According to a Government-General report on industries published in the early 1930s, there were thirty-eight factories in the city of Seoul in 1913. Of these, five mills employed eleven to forty persons, six employed six to ten workers, five hired fewer than five, and family members ran seventeen handicraft workshops.[32] As Sŏng T'aegyŏng and other historians have conveyed, large-scale textile enterprises commenced with the founding of Chōsen Spinning and Weaving factory in Pusan in November 1917. Two years later, in October 1919, the Kyŏngsŏng Spinning and Weaving Company (Kyŏngsŏng pangjik chusik hoesa) built a base in Yŏngdŭngp'o. Although the diverse and manifold procedures of spinning, weaving, knitting, and processing fiber have been overlooked by general observers, cotton textile production required a complex range of jobs. In the first decade of the colonial era, the manufacturing of fiber in Korea did not entail much spinning and weaving because it was simply made into cords. Kyŏngsŏng Textiles, for example, was founded in November 1911 by eighteen investors including Pak Yŏngho and Kim Yŏnsu as the Kyŏngsŏng Cord Company (Kyŏngsŏng chingnyu chusik hoesa). Representative of the domestically oriented, early industrial climate, the Kyŏngsŏng Company at the time produced cords or string (chingnyu) at a time when there was little demand for broadcloth (chingmul). While the company also produced some cloth and hosiery, in its early years the production of cords was ten times that of fabric.[33]

Although the post-WWI era has been highlighted as the inaugural period of large-scale production, textile manufacturing in the 1920s relied more heavily on small- to medium-size mills (chungso kongŏp) and household businesses (kanae sugongŏp). Moreover, because the processing of fiber encompassed a diverse array of tasks, small workshops tended to specialize in one or two particular aspects of production. According to a 1927 Government-General survey, of the 135 textile factories operating in 1924, only fifty-three specialized in the production of woven cloth (chingmulŏp). Two factories specialized in spinning (yŏnsaŏp), four in dyeing and bleaching (yŏmsaegŏp), sixteen companies processed raw silk (saengsaŏp), and sixty workshops concentrated on the initial processing of raw cotton, or cotton ginning (chemyŏnŏp). Large-scale cotton textile plants that incorporated all of the procedures of spinning and weaving were also founded in the late 1910s and 1920s, but their production capacities did not overwhelm smaller businesses. At the near peak of light industrial production, in 1929, approximately one-half of all textiles exchanged in Korea were made in households.[34]

Until the 1930s, the cotton spinning and weaving industry in Korea had a precarious existence due to international competition. The arrival of Japanese spinning giants such as the Tōyō and Kanegafuchi Spinning and Weaving companies in the 1930s drastically altered the climate of the Korean textile sector. Although the first modern factories were founded in the late 1910s, expansion remained slow throughout the 1920s. Between 1930 and 1935, however, the Chōsen Textile Company formed bases in Taegu, Sariwŏn, Ch'innamp'o, Wŏnju, and Taejŏn. Also, the Nanpoku Textile Company, a part of the Mitsui group, installed a plant in Changhang in 1936, and the Kanegafuchi Spinning and Weaving Company established bases in Kwangju in 1935 and in Seoul by December of the following year. In 1933, the Tōyō Spinning and Weaving Company launched plans to build a plant in Inch'ŏn with an approximate four million yen in invested capital. The Inch'ŏn factory commenced operations in 1934, and soon after, in 1937, another mill in Seoul began production. Japanese investment, nonetheless, did not necessarily translate into the Japanese monopolization of textile production on the peninsula. The proliferation of Korean-owned knitting factories in 1930s Pyongyang among other examples confirms that Korean businesses also benefited from accelerated development.[35]

Despite acknowledging the achievements of Japanese capitalists, many scholars in Korea have maintained that the ascendance of modern corporations in Korea was divisive rather than integrative. As alleged by Sang-Chul Suh, large-scale projects "became complementary to Japan's industries" and tended to "produce abrupt disturbances to and detrimental effects on native industries." Although such arguments are plausible inasmuch as the largest companies were Japanese owned, other historians including Carter Eckert and Chu Ikjong assert that colonial policies and practices did not hinder native entrepreneurialism. The fact that large-scale Korean businesses such as Kyŏngsŏng Spinning and Weaving were founded in the late 1910s suggests that the economic boom stimulated Japanese and Korean investment alike. In 1917, thirteen hosiery, or knitting, enterprises that existed in Seoul, including the Kwangmyŏng and Samu companies and Taechang Hosiery, were Korean owned.[36] The number of Korean-owned hosiery factories multiplied throughout the 1920s and even more throughout the 1930s as efficient machinery allowed for greater production. By 1938, there were several dozen more Korean-owned knitting mills in Seoul and sixty-nine in Pyongyang.

Associated with the explosive growth of cotton textile enterprises in knitting, spinning, and weaving was the establishment of coloring (*yŏmsaek*) fabric as an independent sector. The pigmentation division of the Samgong Hosiery Company in Pyongyang built a separate plant for dyeing and bleaching yarns in 1919. In the winter of 1921, the Tongyang Coloring Company (Tongyang yŏmsaek chusik hoesa) was founded in Seoul, and throughout the 1920s, factories concentrating on the coloring of fabric, such as Paekhwa Dyes, and Taedong Dyes, were formed.

The prosperity of Korean-owned businesses in coloring and hosiery production can be explained by the technological characteristics of these industries in the early twentieth century. Manufacturing knitted wear required less sophisticated machinery than cotton spinning, and weaving and was thus more suitable for investors with less start-up capital. Smaller than large-scale spinning and weaving plants, knitting mills generally employed fewer than 100 operatives. Although household manufacturing still accounted for 15 to 20 percent of knitted wear production as a whole, throughout the 1920s and 1930s, hosiery production came to be dominated by medium-scale, Korean-owned enterprises. Similarly, Korean-owned dyeing factories, also operating on less machinery and fewer employees, flourished into the 1940s. The combined effects of technological and economic circumstance helped Korean businesses dominate certain sectors of textile production such as the knitting and dyeing of fabric.[37]

## The Expansion of Female Labor-Intensive Industries

With drastic changes in economic policy after the formation of Manchukuo in 1932, the pace of economic development quickened, reconfiguring the colonial Korean labor force. The focus of production shifted to the heavy industries, but attention to large-scale projects did not necessarily stifle smaller businesses. The 1930s initiated a period of takeoff, or rapid development, when large- and small-scale enterprises prospered through collaboration, thus merging the stages of mechanization. The male workforce, predominantly employed in the heavy industries, experienced "quantitative and qualitative improvements."[38] Whether or not they resulted in betterment, the female labor force also enlarged and evolved. The numbers of female child workers, for instance, increased steadily throughout the period. Girls under fifteen years of age composed 46.1 percent of the child workforce in 1925. This figure surged dramatically to 76.1 percent in 1930, and throughout the 1930s and early 1940s over two-thirds of the child labor force was composed of girls. The sizeable proportion of female child workers, examined alongside women's higher rates of migration to Manchuria and Japan, indicates that the heightened productive demands of the Pacific War reshaped the lives and labors of not only Korean men but women and children.[39]

Filatures and spinning and weaving mills tended to recruit unmarried girls, but other businesses processing food and ceramics, as well as those manufacturing rubber and metal goods, hired older women. As evidenced by the statistical yearbooks of the Government-General of Korea, whereas the proportions of women workers in the textile and food-processing sectors decreased by the 1940s, the percentages of women workers rose in all other sectors including metals, machinery, ceramics, chemicals, lumber, printing, and gas and electricity. The escalating rates of female employment in the chemical and machine tools industries throughout the 1930s reveals that, by the last decade of colonial rule, occupations for women had diversified beyond the realms of textile production.

Because of its technology and scale, much scholarly attention has been paid to the chemical industry and the establishment of mega-scale projects such as the Korean Hydroelectric Corporation (Chōsen suiryoku tenki) in 1926 and the Korean Nitrogen Fertilizer Corporation (Chōsen chissō hiryo) in 1927, funded respectively by Noguchi Jun and the Mitsubishi group. By the late 1930s, these plants represented some of the largest electrical-chemical complexes in East Asia. Also in the late 1930s, other large-scale chemical enterprises, including the Korean Nitrogen Gunpowder Company (Chōsen chissō kayaku) and the Korean Gunpowder Manufacturing Company (Chōsen kayaku seizō), were founded. Nevertheless, most mills producing chemically manufactured goods were not large. The chemical industry (*hwahak kongŏp*) in colonial Korea presented a somewhat strange picture: a small number of giant plants on the one hand and numerous small businesses with "primitive techniques, making fish oil, fish fertilizers," vegetable oils, and rubber footwear on the other.[40] Although corporations making chemical products employed both men and women, females constituted 60 to 70 percent of the total pool of laborers (*chikkong*) in rubber shoe manufacturing projects throughout the colonial era and after Liberation.

While the rubber industry (*komu sangŏp*) was significant in terms of output, a small to medium scale of production characterized it. Pyongyang was the center of rubber manufacturing and some of the larger companies in the area included Sŏjo Rubbers, founded in 1922, Sech'ang Rubbers, formed in 1927, and the Kŭmgang Rubber Company, established in 1929. The production of rubber, which prioritized shoemaking for the majority of the colonial era, shared characteristics with that of knitting inasmuch as many of the businesses were Korean owned. Like hosiery and dyed fabric, rubber shoes (*komusin*) were made in mills with older and simpler technologies. But, because manufacturing footwear required more start-up capital, the rubber industry was unique in its tendency toward joint ownership (*hapja*). Although in 1924 the two businessmen Kim Tongwŏn and Yi Yŏngha established P'yŏngan Rubbers (P'yŏngan komu), rubber mills were more often financed by several investors. The Tonga and Ch'ungch'ang Rubber companies were each founded by five men in 1921 and 1922. Similarly, six entrepreneurs formed the Taedong Rubber Company in 1923, and seven businessmen founded the Tongyang Rubber Company in 1926. The Sŏgyŏng Rubber Company, established in 1925, was initiated by the joint effort of eleven investors.[41]

From its emergence in the late 1910s and 1920s, rubber manufacturing expanded throughout the colonial era, reaching its height in the 1930s.[42] The production value of rubber footwear rose from 10.5 million yen in 1935 to 17 million yen in 1937—a 62 percent growth in the span of two years. Aside from statistical evidence, newspaper reports also demonstrate how rubber shoemaking companies were able to remain competitive over such a lengthy period of time. As described by an editorial in the *Sŏjo ilbo* (Western Korea Daily), whereas export-processing sectors suffered from the reverberations of the global Depression, the rubber industry that relied on the demands of rural and local society was able to prosper throughout the 1930s.[43]

Predictably, Korean women comprised a small percentage of workers in heavy industrial projects such as mining, but their discernable rise in numbers during the late 1930s and early 1940s illustrates the changing characteristics of women's work in the latter years of colonial rule. According to a 1933 Government-General survey, approximately 3 percent of the mining (*kwangŏp*) labor force was female. By 1941, their composition more than doubled to 7.3 percent and continued to increase throughout the early 1940s. Female miners usually worked outside of the shafts in the collieries of Hwanghae and South Hamgyŏng provinces, which specialized in coal (sŏktan) and pig iron (sŏnchŏl) mining. In coal mines, women screened and sorted coal and stacked it to prepare for coking. In iron mines, women removed cinders from furnaces and chipped bits of iron ore with small picks. They also took iron ore from trams, sorted out the stone and shale, and cleaned the ore. By 1941, the restriction on women working inside the pit (kaengnae) was lifted, and women started strenuous work inside the shafts. Although the number of female employees in collieries did not exceed 50,000, their entry into heavy industrial work confirms that the varieties of women's industrial labor altered considerably by the 1940s.[44]


*Summation*

One of the most popular perceptions of modernization was its adverse impact on the traditional family system. Each stage of industrial differentiation and specialization "struck at the family economy, disturbing customary relations between man and wife, parents and children," distinguishing more sharply the boundaries between work and life. Meanwhile the family was roughly "torn apart each morning by the factory bell."[45] Whereas premodern modes of production were centered on households, as families became wage earning instead of producing units, their members no longer shared common interests in the property guaranteeing their livelihood. In their analyses of how mechanization influenced female labor, some scholars have asserted that although women's productivity was high in the pre-industrial household economy, it declined in industrializing economies.

Late twentieth-century historians of women's labor, including Scott and Tilly, inspired groundbreaking perspectives of the history of women's work. Nonetheless, some of their conclusions reaffirm the ideas of Marxist feminists who maintained that "women became more dependent upon the employer or labor market" as they looked back on a "golden past" of autonomy in household enterprises.[46] Such evaluations rightly underscore the fact that women in traditional societies held the reigns over family funds and their allocation. Still, overemphasizing women's agency in the arenas of premodern domesticity suggests that industrialization and modernization undermined this authority. A closer examination of the Korean case, however, shows that domestic and nondomestic accomplishments were not necessarily antagonistic.

To understand the rationales of working women in colonial Korea requires placing these individuals within peasant households and close-knit rural communities. The dynamics of these communities, however, have been hotly contested by social scientists. Their discussions generally focus on peasant consciousness and analyze the economy rather than culture, but they also outline the possible impulses for workers' activism. Based on a moral interpretation, first expressed by James Scott, peasants who were traditionally constrained by the "vagaries of weather and tributes in cash, labor, and kind" sought "safety first" before seeking profit or material accumulation. Other scholars, such as Samuel Popkin, challenge notions of rural conservatism by asserting that though peasants enjoyed stability, they sometimes compromised security for future rewards and that they were "forward-looking" planners, rather than those who solely embraced subsistence-based lifestyles. Still others contend that wealthier farmers acted for political reasons whereas poorer peasants were morally driven. Despite the intricacies of these debates, the conspicuous rise of female industrial labor indicates that the majority of rural families in early twentieth-century Korea maintained stability *and* took risks in their everyday ventures.[47] For unmarried peasant girls, the implications of their wage work were wide ranging. When working daughters lived at home, their wages unquestionably became a part of the family fund. Even if working away from home, some young women saved their earnings to give to their parents, thus exhibiting the tenacity of tradition in a wage-economy setting. Whether they sent their wages back home or not, if living elsewhere, they still fulfilled their functions in the family economy. Simply by moving into a factory, a daughter aided the continuation of the household economy by relieving her family of an extra burden.

Women workers of the colonial era have been depicted as moral agents who were quiescent actors moved by security-based impulses. A closer reading of the many motives for factory work, however, indicates that though women might have joined the workforce to appease parental pressures, they were simultaneously planning for their future welfare. In a time and place in which a woman's greatest ambition in life was marriage, a dowry was her only chance for social mobility. Most unmarried working women in early twentieth-century Korea could not depend on networks of kin to provide for their future welfare and therefore relied on themselves. Just as nineteenth-century commoner girls became domestic servants to compensate for inadequate family funds, women took on factory labor to strengthen their financial position and improve their marriageability. Although their aspirations for work were drawn from tradition, the ways in which young women accomplished these aims transformed by the latter years of the colonial era.

## THREE

## Lives and Labors Inside the Factories

It was an especially cold winter afternoon, several days after the lunar New Year in 1944. In a village near Pusan in South Kyŏngsang Province, two men called on the Yi family residence. They offered to the youngest daughter of the household, Chungnye, education and employment at the Kanegafuchi Spinning and Weaving Company in Kwangju, some 200 kilometers west of Pusan. For Chungnye's parents, the men provided relief at a time when they could not care for the entire family. For Chungnye, the recruiters promised valuable training and improved living conditions. The following day, Chungnye, for the first time in her life, left her childhood home and began her journey to Kwangju.

Like many of her contemporaries, Yi Chungnye entered into wage work with parental encouragement. For others such as Kim Chŏngmin, however, factory work presented a break from the bounds of parental authority. Kim denied that her parents' approval or disapproval of factory work would have altered her decision. Kim Chŏngmin's sentiments of her earliest years of work contrasted with those of Kim Ŭnnye. Born and raised in Pyongyang, Kim Ŭnnye suffered material and emotional difficulties from an early age because of her mother's premature death and her father's alcoholism. Kim says family circumstances, coupled with the poverty rampant in South P'yŏngan province, gave her little choice.[1] Whether beguiled by empty promises or pressured by household needs, once women reached the silk-reeling, spinning, weaving, knitting, and rubber mills, they encountered settings wholly unlike those described by recruiters. There, they worked for twelve to fourteen hours a day. Residence within the factory compounds was often customary. In colonial Korea as in Japan, many factories strictly prohibited their employees from "leaving the premises, visiting home and writing and receiving letters."[2]

Viewed from the pinnacle of high politics, the distinctions of Korean industrialization appear to stem from the multiple strategies of colonial enterprise. Japanese businesses with near fifty years of corporate experience installed in Korea a management system that used strategies of surveillance and discipline, which separated, analyzed, and differentiated individuals. Through devices that Michel Foucault calls "schemata of constraint," including "time-tables, compulsory movements, regular activities, solitary meditation, and work in common," new, social scientific methods of human organization created, subjected, and practiced "docile" bodies. In factories, discipline was achieved through the distribution of bodies; "the spatial arrangement of production machinery and the different forms of activity in the distribution of 'posts' had to be linked together." Within these posts, or ranks, determined by location, time, and context, activity was controlled. By establishing and scheduling rhythms and techniques to workers' interaction with the machinery, owners optimized output with efficiency. Each variable of the labor force—strength, promptness, skill, and consistency—was observed, characterizing, assessing, computing, and relating "to the individual who was its particular agent."[3]

Viewed from the pockets of everyday life, however, modern economic development in Korea depended on the support or resistance of individuals. Although the colonial state and Japanese capital influenced the experiences of working women, so did their immediate communities and families. Systems of labor management in East Asia, as Ching Kwan Lee highlights, were bolstered by "familial hegemony," or corporate ideologies and behaviors that drew on Confucian social doctrines.[4] Still, it was not only the factories that employed the model of the "family," as real families propelled the activities of factory operatives. Model employee behavior did not necessarily signify strong labor relations, so long as workers' motivations mostly came not from within the factory but from without.

The fact that industrial laborers were also family members gains greater significance when examined alongside theories of "strategy" versus "tactic" as articulated by Michel de Certeau and John Law. As these and other scholars suggest, whereas ordinary people seize benefits through tactics, they are unable to affect long-term strategies. According to de Certeau, strategies are the "calculus of force-relationships which become possible when a subject of will and power," a proprietor, a government, or an institution, can be isolated from others. Strategies, such as shop floor discipline, executed in fixed spaces allowing for the repetition of hegemonic practices, are institutional and structural. Tactics, on the other hand, are a "calculus which cannot count on" a proper location; "the place of tactic belongs to the other." Reliant instead on time, tactics require turning quickly moving situations into opportunities.[5] Factory women in colonial Korea undeniably accommodated and resisted corporate strategies with tactics, or calculated their actions, taking advantage of fortuitous occasions within hegemonic spaces. Nevertheless, to limit the agency of workers to the tactical overlooks their far-reaching aims. Although strategies cannot be derived simply by "adding up tactics," factory women, as members of households, acted for the strategic interests of their families as defined against competitors or adversaries of these units.[6] Employee performance was not exclusively affected by management, as discontented factory workers often continued to toil for their kin.

While parental authority might have declined with industrialization, most unmarried women in colonial Korea initiated wage work to fulfill the demands of their parents' households. "Leaving home, getting married, setting up independent households and all such individual transitions were related to collective family decisions," affecting directly and indirectly the household structure. Therefore, individual commitments such as commencement of work and the "movement of different family members in and out of the household" were synchronized with the motions of the family as a collective unit.[7]

Whether parents encouraged or permitted their daughters' departures, the adoption of wage work relied on the integration of individual "life-plans" with the plans of the family and household. As Tamara Hareven explains, "life-plans" encompass a broad spectrum of "goals and aspirations around which an individual or family organizes its life." Defensive plans focus on coping with "recurring crises and insecurities," and long-range plans, "spanning two or three generations," are to assure security and achieve long-term advancement. Although Korean women might have entered the factory to satisfy the defensive needs of the family, many former factory girls later achieved long-range plans including higher education and surpassed the economic security held by their parents and grandparents. For young rural women of colonial Korea, wage work was the first step in assuming their familial and social responsibilities.[8]

Disciplinary devices, such as "cells," "places," and "ranks," were enacted in intricate spaces that were architectural, functional, and hierarchical. Under constant surveillance, these spaces, roles, and procedures cultivated and regulated docile, efficient bodies. Under limited surveillance, however, these same "cells," "places," and "ranks" enabled workers' tactics of foot-dragging, feigned accommodation, and strategies of resistance. Since supervisors, however thorough, could not monitor all of their employees all of the time, in the workshops and dormitories of colonial Korea, there was neither an exact convergence nor a profound opposition between what could be termed "legal" and "illegal" behavior. Communities formed on shop floors and in residence halls maintained "relations with one another that involved not only rivalry, competition, and conflicts of interest, but also mutual help and complicity." As Foucault implies, in any given situation of power the "reciprocal interplay of illegalities" was just as pervasive as the operations of legality.[9] Female factory workers in colonial Korea, therefore, were able to use the "laws, practices, and representations that were imposed on them," subverting them from within.[10] Nonetheless, it is important to remember that working women, allied with their families and their fellows in the workshops and dormitories, not only resisted the abuses of capitalism but also contributed to the continuation of colonial modernization.

To assess the myriad of ways in which Korean women became industrial workers requires a discussion of the structural and familial forces that helped configure the efforts and experiences of factory women. Thus, the first half of this chapter concentrates on the procedures of manufacturing that contoured the dynamics of female industrial labor. I outline the productive processes of female labor-intensive enterprises that emerged and expanded in the 1920s and 1930s: sericulture and silk reeling, cotton spinning, weaving and knitting, and rubber shoemaking. Although cotton textile manufacturing, silk-reeling, rubber shoemaking, and food-processing plants hired mostly women, the standards of work in these factories and their labor pools varied by industry. Sericulture was done on silkworm farms, predominantly in south-central regions such as the Ch'ungch'ŏng provinces. Silk-reeling mills, owned by both Korean and Japanese businessmen, were located in both large and small cities close to the farms and were usually medium in scale. While there were numerous small cotton spinning and weaving enterprises, the best documented were the Japanese-owned, large-scale factories that built their bases on the peninsula's main urban centers. Knitting factories, on the other hand, were smaller projects that were often Korean owned. Similarly, rubber mills, generally centered in big cities such as Pyongyang and Pusan, were small to medium-sized businesses owned by both Korean and Japanese investors. Nevertheless, the nationality of the owners did not affect working environments to the extent of industrially specific organization, scale, and technology. These structural distinctions between female labor-intensive industries determined the methods of recruitment, payment, and conditions of women's work in unique ways.

The working lives of women in colonial Korea were inevitably mediated by economic and manufacturing environments, but they were also shaped by their families and their familiar relations in the factories. Although before the 1920s, women acquired factory jobs through relational or community-based connections, by the mid-1920s, direct recruitment became more common. The operatives of food-processing and rubber mills, who were frequently older, married women, could work seasonally without lengthy terms. The usually younger, unmarried workers in silk reeling and cotton textile manufacturing, however, were contracted for three to five years. Their wages often went directly to the parents, illustrating the extent of female obligations in household economies. For these young women, workdays were interminable; cotton and rubber manufacturers generally endured twelve-hour shifts whereas those in silk reeling worked for up to fourteen hours daily. For those employed in textile mills, the other half of the day was spent in dormitories. The tradition of Confucian paternalism buttressed in-factory residence, but its application was not necessarily gender specific because many male youths were also housed by companies. Despite the shortcomings of the corporate residence system, dormitories served as arenas of commiseration, friendship, and more instrumentally, as breeding grounds for dissent, as night work, accidents, disease, death, and increasing corporate control incited women to act.

Examining social change from the perspectives of workers reveals that the customs and accommodations of the native populace paved a region's path to modernization as much as authority from above. Foreign encroachment, colonization, and civil war were only a few of the unhappy landmarks that Koreans encountered on the path to industrialization. Throughout these progressions, the actions of the populace, whether cooperative or resistant, also contributed to economic development. How young women entered the mills and their conditions of work and life demonstrate that the forms of compliance and contestation factory women expressed not only influenced the making of their life courses and identities, but also aided in the construction of the early twentieth-century Korean economy and society.

### The Lines of Production: Silk Reeling, Cotton Textiles, and Rubber

In traditional Korean society, the techniques of sericulture were reserved to upper-class craftswomen. By the 1920s, however, silk breeding, reeling, and weaving became commercialized. Like cotton spinners and weavers, silk reelers have been granted much historical attention as disenfranchised, female, colonial subjects. Little consideration, however, has been given to the fact that in addition to the factory labor force, silkworm farmers were often women who suffered the many abuses of modern agricultural wage work. A review of the procedures of silk manufacturing sheds light on women's dual roles as agricultural and industrial workers.

Raw silk of the finest quality came from the common domesticated silkworm, *Bombyx mori* (*cham* or *nue*), which fed on the leaves of mulberry (*sang*) trees. Sericulture (*yangjamŏp* or *chamŏp*) in colonial Korea drew upon an age-old tradition of cultivating caterpillars from the egg stage through to the completion of the cocoons. The making of silk began with the selection of silkworm moths (*chama* or *nue nabang*) for breeding, and the egg-laying process was supervised. Breeders sold the eggs to silkworm farmers (*yangjam nongga*), who then hatched the eggs. Apart from their rearing, silkworm farming entailed the cultivation of the mulberry trees that provided the leaves upon which the silkworms fed. Thus, these farmers prepared the nursery (*myosang*) by cultivating the mulberry trees (*sangmyo*) and managed the mulberry fields (*sangjŏn kyŏngyŏn*).[11] Cocoons (*chamgyŏn* or *nue koch'i*) constructed by silkworms were transferred to filatures, where they were reeled, spun, and woven into cloth.

The spinning of silk took place in filatures, or silk-reeling mills (*chesa kongjang*). Before silk threads were drawn, the dried cocoon (*kŏn'gyŏn*) was placed in hot water to soften the sericin, a substance that bound the silk threads. After soaking, silk fibers from the cocoons were unwound or reeled (*sosa*). Depending on the thickness of the thread desired, a few or several cocoons were unwound simultaneously. Silk-reeling machines (*sogi* or *soch'a*) extracted extra moisture from the strands and, to help the sericin hold the fibers better, filaments were crossed in a *croisure* pattern. The strands were then wound onto reels and dried. After reeling, they were twisted together to form stronger threads through a procedure known as throwing (*kyŏngyŏng*). By boiling the filaments in an alkaline solution, the threads were then degummed, or extracted of sericin. To regain density, the silk was weighted (*ch'ingsa*) through the application of a finishing substance. Finally, the threads were woven (*kyŏnjik*) and dyed (*yŏmsaek*), inspected (*kŏmsa*), and packaged (*p'ojang*) for distribution.[12] Despite the range of tasks in filatures, because of the industry's reliance on rural silk farms and manufacturing methods that did not operate on twelve-hour shifts, silk reelers suffered through some of the longest hours of work among contemporary factory women.

Although mechanization transformed the spinning and weaving industry, modern procedures of drawing thread from cotton, much like silk reeling, were based on timeless customs. Before raw cotton (*wŏnmyŏn*) was spun, cotton fibers were cleaned through scutching (*t'amyŏn*) or cotton ginning, which consisted of detaching fibers from stems by beating the raw cotton to shake seedpods and other components loose. After removing impurities, an assortment of different kinds of fibers were then combined with the cotton through mixing (*honmyŏn*). Workers scutched and mixed (*hont'amyŏn*) by feeding masses of fibers into machines, which separated them into thinner cords.[13]

After the raw cotton was scutched and mixed, it was further refined through carding. In the carding room, workers fed the scutched and mixed cotton into massive machines fitted with rows of wire teeth, used to disentangle the cords to form a sliver, or a continuous strand of loose fiber. Because of the dangerous machinery, carding room operatives were the most prone to hazardous accidents. The carded cotton (*somyŏn*) was then combined and pulled into uniform thread through drawing (*yŏnjo*). Upon drawing, the thread was pulled into a finer string through coarse spinning, or roving (*chobang*), and then spun into a thinner strand through fine spinning (*chŏngbang*). Ring frames were used in the final stage of spinning in which the thread was wound (*kwŏnsa*), doubled (*hapsa*), twisted into a continuous strand of yarn, and reeled (*injo*). Following inspection, it was bundled (*okch'e*) and packaged. The production of cotton yarn (*myŏnsa*) in colonial Korea was thus a twelve-stage process involving a complex of procedures.

Cotton spinning, weaving, knitting, silk reeling, and rubber shoemaking necessitated a multitude of tasks. In spinning mills, for instance, workers could be assigned to any of the following divisions: mixing and scutching, carding, drawing, roving, spinning, as well as carpentry, mechanics, and electronics. Even in the early phases of mechanization, women were employed in the most labor-intensive divisions: sections for the drawing of sliver (*yŏnjo*), coarse spinning or roving (*chobang*), fine spinning (*chŏngbang*), and reeling (*injo*). Novices in spinning mills usually started work as piecers, joining the ends of yarns that broke due to tension and temperature as they emerged from the machinery and wound onto bobbins. Piecing required great speed, dexterity, and persistent concentration. Although the least paid, piecing was one of the most exhausting jobs, and one of the most dangerous, with many women losing fingers drawn into machinery.[14] Female employees dominated the cotton-spinning industry, and by the last years of the colonial era, spinning factories were almost entirely staffed by women.[15]

The automation of cotton textile machinery greatly advanced the technology of weaving.[16] Whereas before 1926, the Korean textile industry was focused on spinning cotton (*pangjŏk*) and silk reeling (*chesa*), by the late 1920s, these factories installed automated looms, and new spinning companies established weaving divisions. Before weaving, the warp thread (*kyŏngsa*) was set onto the loom through a procedure called warping (*chŏnggyŏng*), which was followed by sizing (*hobu*), or starching, and reeding.[17] Automated looms wove the threads (*chikp'o*), but machines did not remove the need for human labor. In weaving rooms workers performed the preparatory chores of warping (*chŏngkyŏng*) and setting the warp thread (*kyŏngsa*), and operated the looms. They starched the warp threads and separated the threads onto reeds, or movable frames. Just as yarns broke as they emerged from spindles, depending on the quality of the loom, its age and maintenance, they would break during weaving. Whenever a weft yarn (*wisa*) broke, piecers would tie the yarn at the break and continue the operation. Women also cleaned the weaving area and the machines. After the cloth was woven, it was taken to the finishing room, inspected, and packaged into rolls (*ch'ŏnmungch'i*).

Weaving, however, was not the only method through which fiber was made into fabric. Just as women wove apparel goods, fibers were compressed to create felt, or braided to make lace or knitted goods. In weaving, taut threads were interlaced, but in knitting (*meriasŭ*), fabric was produced by interlocking loops of yarn. Weft knits (*wip'yŏn* or *wi meriasŭ*) were made

*Plate 3.1*    Cotton ginning or "scutching and mixing" workshop (1920s).

*Plate 3.2*    Novices piecing thread in a spinning room (1920s).

from one continuous piece of yarn, generating an extremely elastic fabric used to make stockings, socks, sweaters, and undergarments.[18] In warp knits (*kyŏngp'yŏn* or *kyŏng meriasŭ*), however, many yarns were used and the yarns formed zigzag patterns so that different yarns formed loops directly above one another. Whereas weft knits were tubular, warp knits were flat, and in colonial Korea, the manufacturing of weft knits, particularly socks and undergarments, was most prominent.

For much of the early twentieth century, hosiery factories were small in scale and involved the spinning, knitting, and dyeing of raw cotton into knitted goods. Although the knitting industry never attained the scale of cotton spinning and weaving, the processing of knitwear became more specialized in the late 1930s. Similarly, in the early 1920s, fabric was dyed in departments (*yŏmsaekbu*), whereas by the late 1920s, the coloring of fabric (*yŏmsaek*) was done in factories exclusively dedicated to dyeing.[19] Notwithstanding that the procedures of coloring fabric and manufacturing hosiery varied from those of cotton spinning, weaving, and silk reeling, knitting and dyeing plants also preferred to hire young women.

While female employment in the chemical industry during early development appears somewhat misplaced, the tasks of rubber workers, who primarily manufactured shoes, were arguably less technical than the labors of their counterparts in textile factories.[20] Shoes were made by craftswomen who sewed or glued together separate parts of the shoe, which formed a prototype. Although the degree of departmentalization was determined by factory size, rubber footwear production in colonial Korea involved five general procedures. Before rubber was made into shoes, stocks of crude rubber, imported from Southeast Asia, India, and China, were chemically and physically masticated, or shredded, and mixed with fillers. Mastication, or the initial mixing and blending of rubber, took place within the mixing and blending division (*paehapbu*), where bales of crude rubber were cut into pieces and peptizers were added to soften the rubber. The actual shredding was done inside mixing machines that both separated and blended the compounding ingredients. Upon mastication, the rubber was flattened into sheets in the rolling division (*nolbu*).

The two operations preceding the actual cutting and piecing of shoe segments required the strength to move large containers of chemicals and rubbers. These divisions employed male workers to haul the material, whereas machines masticated and rolled the rubber. The shoemaking portions of production, on the other hand, relied heavily on the dexterity of women workers. In 1935, approximately 73 percent of the manufacturers in the shoemaking, workshop divisions (*changbu*) were women.[21] In these rooms, pieces of rubber were cut into the various parts of shoes, such as uppers, outer soles, insoles, and heels. The sections were then glued together and set into aluminum molds. Using their hands and simple tools, workers further shaped the rubber into molds, using oils to smooth the melding process. The shoes were then inspected, and if the finished product was unsatisfactory, they were returned to the worker for reconstruction.

After shoes passed inspection, they were taken to the vulcanizing division (*karyubu*). Vulcanization is the process through which rubber gains density, loses its adhesive quality, and becomes more resistant to water. Although vulcanization entailed the simple procedure of heating rubber with sulfur, additional steps for strength and durability were added as technology advanced. Accelerators such as aniline, which increased the rate of vulcanization, were frequently mixed with sulfur. To further insulate the rubber, activators such as zinc oxide were also used. Following vulcanization, the shoes were transferred to the finishing room (*mamuribu*). They were taken out of molds, refined by grinding away rough edges, and packaged.

Although rubber mills hired women, the characteristics of work for employees in the rubber industry were quite dissimilar to those in textiles. Unlike spinners, weavers, and silk reelers who were recruited from the countryside and housed on company grounds, rubber factories employed day laborers. Because shoemaking did not rely on continuous machinery or highly rationalized systems of operation, it was commonly done by married women who could work with their children on the shop floor. Shoemakers also worked seasonally if their productive demands conflicted with their reproductive and domestic responsibilities. With little need for young recruits from far away, rubber factories had few dormitories and less comprehensive forms of labor control. Nonetheless, the employees of both rubber and textile manufacturing companies shared the distresses of inadequate conditions and wages.

*Recruitment, Contracts, and Wages*

Whether factory workers were recruited by company agents or introduced by relatives depended not only on the developmental milieu but also on the industry of their employment. By the 1930s, most spinning and weaving firms in Korea recruited their employees directly. But workers in the more tradition-bound silk-reeling projects entered the mills through familiar means. A rare and comprehensive study of the silk-reeling industry by Yuk Chisu, published in 1938, shows that 905 of the 1,929 employees of the Kanegafuchi (Chongyŏn) Silk Reeling factory in Kwangju acquired their jobs through relatives. The fact that 46.3 percent of its employees found work through networks of kin, whereas only 16.9 percent of the labor pool was recruited, indicates that the mobilization of silk reelers relied on local traditions. Young women also sought work at the factory voluntarily. About a quarter of the Kanegafuchi filature's workforce, 25.6 percent, entered the mill by responding to advertisements, suggesting that more women in silk reeling, as opposed to cotton spinning and weaving, gained employment through direct application.[22]

Situated near silk farms, filatures retained older methods of labor mobilization, whereas the management procedures of large-scale cotton textile plants were directly influenced by the accelerated economic development that was under way by the late 1920s. Colonial labor policies that sought to transfer (*idong*) workers across provincial boundaries corresponded with corporate plans for social engineering. The Government-General's 1930 Policy on the Regional Regulation of Labor Supply (Chiyokkan nomu sugŭp chojŏngchaek), for example, aimed to redistribute the Korean workforce. Among those affected by this plan were female textile operatives who, throughout the 1930s, moved further and further away from their homes. For state officials, the domestic and regional migration of labor facilitated faster industrial expansion. For mill owners, workers' distance from home abetted management's control.

Methods of recruitment (*mojip chedo*) developed in stages, according to industry and locale. Just as landlords of premodern estates depended on stewards to obtain new servants, executives during the first years of modernization relied on foremen, supervisors, and other established employees to procure their workers.[23] As the technologies of manufacturing evolved and the demand for labor increased throughout the 1920s, large textile companies hired recruiters. Unlike the *oyakata* of early industrial Japan or the "coolie head" of China, the intermediary between the Korean worker and the firm was not a contractor (*ch'ŏngbuwŏn*) but a recruiter (*mojibin*). By the 1930s, companies established recruitment offices in the provinces to find young women and deliver them to the mills.

Most of these women entered factories with little or no urban work experience and fewer employable skills. Although contracts for manual laborers in silk-reeling, knitting, spinning, and weaving workshops in the 1910s may have been orally transmitted, by the 1920s, printed contracts were prevalent. The promise of fixed pay and bonuses led women to the mills. But, once inside the factory, they found that their wages were anything but fixed and bonuses were far less frequent than penalties. For terms of three to five years, employees agreed to remain on the factory and dormitory premises, serving the company physically through labor and mindfully through obedience. The owner set the salaries, which were contingent on whether the women fulfilled their duties, adhered to the company regulations, and met production targets. During a time when few peasants could read their contracts, much less understand them, workers often allowed the employer to dock wages arbitrarily and acknowledged the company's unilateral right to dismiss them at any time for reasons it deemed appropriate.[24]

Wage work offered assurances of fairness and job security, but for women workers of the 1920s and early 1930s, salaries were the primary tools with which human resources were regulated and restrained. A definitive characteristic of textile labor was the production-based wage system (*togŭp chedo*) that measured wages according to the quality and quantity of the goods made. Whereas unskilled workers in textile mills generally earned production-based wages, those classified as employees (*chigwŏn*), or skilled workers, received fixed salaries according to a daily wage system (*ilgŭp chedo*). Nevertheless, most factory women received payment according to their fulfillment of the production requirements. In Korea, although they inspected finished material, supervisors did not directly allocate employees' salaries. Therefore, the industrial wages of Korean women were less secure than those of Korean men or workers in Japan but more standardized than those of women in Chinese cities such as Shanghai.

In addition to production-based wages, female labor-intensive industries upheld policies that undermined the earnings guaranteed by contractual agreement. Production-based payment already dampened workers' security, but other retributive methods, such as penalty fees (*pŏlgŭm*) and the forced compensation of inferior goods (*pullyangp'um paesang*), also detracted from actual earnings. Such penalties were especially detested by workers because they let the company blame and charge them for the factory's faulty tools and machinery.[25] In many mills, including those hiring women, portions of workers' salaries were withdrawn automatically as "compulsory savings" (*kangje chŏgŭm* or *kangje chŏch'uk*). Even if employees fulfilled the terms of their contracts, these savings frequently went unpaid.

Ethnically biased wage standards have been ascribed as the primary reasons for labor activism. For sure, Koreans faced discrimination, as Japanese workers in Korea earned twice as much as Koreans and men earned more than women. These general wage scales have been used to illustrate the severity of Japanese exploitation.[26] Nonetheless, the argument that labor protests were galvanized by unequal payment or treatment loses credibility when recalling that throughout the colonial era the Japanese composed less than 3 percent of the peninsula's population. A small community comprised mostly of bureaucrats and opportunists, a mere 17 percent of these settlers engaged in industrial labor. A critical and comparative evaluation of pay scales in construction, manufacturing, and mining indicates that general salary guidelines did not mean that wages were the same for those of the same sex or nationality. Ethnicity and gender imposed ceilings, but conditions also varied for child workers, and earnings also fluctuated according to demographic, economic, and technological trends.

Popular understandings of wage differentiation in the colonial era presume competition between Koreans and Japanese, as well as men and women, but most employees worked with others of their own age, sex, and ethnicity. The labor force of each industry was relatively sociologically homogeneous throughout the colonial era, and Korean women rarely competed with men or the Japanese. Japanese men, for instance, received the highest pay in the machine and machine tools sector that hired many Japanese technicians but earned the least in textile factories that employed few Japanese. Both Korean and Japanese women earned almost as much as men in food- and chemical-processing, shoemaking, and textile projects, where women were most prominent. Because few females, Korean or Japanese, were employed in heavy industries until the last years of the colonial era, the wage scale across industries worked in their favor as women earned the most in light industries. Minors, however, were less fortunate. Textile manufacturing, which predominantly hired youth, was the lowest-paying industry for child workers of those recorded.

The high proportions of female and child workers have long been interpreted as examples of colonial exploitation. Nevertheless, these arguments lose their credibility when viewed with the numbers of women and children in the Japanese workforce. A comparative survey of contemporary statistics reveals that compositions of workers by age, sex, and sectors were remarkably similar in Japan and Korea. Throughout most of the 1920s and 1930s, both Korean and Japanese women comprised more than 30 percent of their total industrial workforces.[27] Likewise, in 1938, child workers composed 9 percent of the labor force in Japan, whereas this figure was closer to 10 percent in Korea. Wages for women and children were similarly meager for workers in Japan. Still, Korean and Japanese female and child workers were distinguished by perhaps more pressing factors than wage or employment opportunity; namely, the conditions of labor they encountered in the factory.

## *Work Inside the Factory*

Young female mill operatives in early twentieth-century Korea generally resided in dormitories, but many others, especially married women, commuted. Company residence was widespread, but the industry of employment, factory size, and regional customs also determined habitation; day labor was more common in big cities such as Seoul, Pusan, and Pyongyang, whereas factories in smaller cities more often housed their employees. The employees of medium to large-scale textile firms, for example, lived in company housing, whereas factories specializing in knitting, dyeing, and small-scale spinning and weaving, as well as rubber, tobacco, and food processing employed day laborers. The procedures of recruitment and the terms of labor in colonial Korea were diverse, but routines inside the factories were similarly arduous.

A worker's day began at dawn, but some might have risen earlier, if they lived far from the mill. Kim Ŭnnye, for example, recalled waking with her grandmother when she heard the crows of roosters. Although Kim lived outside of Pyongyang, those living in cities awoke to the sounds of mill whistles, which first blew at 4:30 A.M., waking in-factory residents. Working women, if they lived at home, were most likely expected to fulfill domestic responsibilities. Before leaving for the mill, women drew water for the day from wells or nearby creeks and prepared morning meals. For the poor, breakfast usually consisted of grains such as barley; if fortunate, rice was mixed with grain and accompanied by pickled side dishes, consisting of cabbage, turnips, or bean sprouts.

Morning chores were especially difficult for the young women who had to commute long distances. Kim's journey with neighbors from her home to the factory took over an hour.[28] Despite the constraints of dormitory life, one advantage of in-factory residence might have been the boarders' proximity to the workplace, given the difficulties of commuting, which could have involved traveling for hours daily to and from work. The adversities of travel and the vagaries of weather did not assuage angered supervisors, who checked each morning for attendance and promptness. The whistles blew again at a quarter to six, and the day laborers who had been waiting outside the gate entered, many "half-asleep, yawning and coughing."[29] In cotton textile mills, the day shift officially began at 6:00 A.M. but workers had to enter the factories earlier so that they would be at their stations when the night shift worker finished, thereby keeping the rhythms of the machines in sync. After attendance was checked, women changed into work clothes, most likely aprons with pockets to hold bobbin cases, scissors, and other tools. At exactly 6:00 A.M. another whistle blew, signaling the end of the workday for those on the night shift and the start of the day shift.

Once inside the factory, workers faced drastic transitions in environmental conditions. The temperature inside cotton textile workshops, around 15°C to 20°C in December and January, might have provided immediate relief for commuters in the winter, but in the summer, the temperature inside the workshops often rose above 38°C. The relative humidity, rising up to 100 percent during the summer months, also caused discomfort. Apart from the heat, cotton dust was a permanent environmental hazard. It was everywhere, in weaving and knitting rooms, but most pervasive in spinning workshops. In the various divisions for drawing, roving, and spinning, particles of cotton visible to the eye emitted from the machines and circulated in the air. Although workers constantly swept between rows of machines, cotton fluff drifted on the floor, covering it in moments after it was swept. In the scutching and mixing rooms (*hont'amyŏnbu*), the dust was, of course, worse. Operatives in cotton textile manufacturing frequently worked covered with white fluff, which entered the eyes, nose, mouth, and ears and stuck to hair and eyebrows. In the midst of intense heat, humidity, contaminated air, and poor ventilation was unbearable cacophony. In spinning rooms, rows upon rows of machines roared so loudly that it was impossible for workers to hear each other unless they shouted. In weaving rooms, the din intensified with the sound of the looms' shuttles in constant, cantankerous motion.[30] The conditions of work in cotton textile industries might have been no worse than those of male labor-intensive projects, such as coal mining—but women and children worked longer shifts than men.

According to a 1933 Government-General report on factories and mines, the industries most reliant on female and child labor demanded the longest hours of work.[31] The study shows that out of the female labor-intensive sectors examined, rubber shoemakers enjoyed the shortest workdays, from eight to ten hours, whereas women in spinning, weaving, and knitting worked in twelve-hour shifts. By the 1930s, most factories producing cotton textiles operated on machinery that ran twenty-four hours a day, and reduced their shifts to twelve hours. But in silk-reeling mills that did not rely on continuous machinery, thirteen- to fourteen-hour workdays were common. The Kanegafuchi Silk Reeling Factory (Chongyŏn chesa kongjang) in Kwangju, for example, operated in thirteen-hour shifts, as did Taech'ang Textiles (Taech'ang chingmul) and Hansŏng Textiles (Hansŏng chemyŏn) in Seoul.[32] Some companies, such as Katakura (P'yŏngch'ang chesa) and Kunze Silk Reeling (Kunsi chesa), upheld fourteen-hour workdays, which became a central issue of contention in the 1931 strike of the Katakura Silk Reeling workers in Yŏnghŭng. Analyzing shop floor standards across industries indicates that women employed in silk-reeling enterprises endured the longest hours of work.

On shift, these operatives worked almost without pause. As recorded in the 1933 report, seventy-seven of the 140 textile factories (65.8 percent) surveyed provided sixty to ninety minutes per shift for rest. Although most mills allowed for at least sixty minutes of rest, twenty-five factories (15.2 percent) offered only thirty to sixty minutes.[33] Recess, however, was not necessarily a period of rest but meant time spent away from the station. Mealtimes were included in the breaks; most women took thirty to forty minutes for their two meals over their twelve-hour shifts. Thus, if they had twenty to thirty minutes remaining for respite, workers spent this time cleaning the machinery and their stations. The duration of breaks fluctuated according to industry and factory as well as employee positions in the corporate hierarchy. Technicians and supervisors enjoyed the longest periods of recess, and male laborers took regular cigarette breaks, but most young Korean women, usually in the lowest ranks, had no such privileges.[34]

Although a diverse array of jobs was available to women in cotton textiles, unskilled recruits were first assigned to the spinning mill. Novices were most prevalent in roving and spinning rooms, which housed three types of manual laborers: doffers, piecers, and sweepers. There, the majority of young women were piecers. The particularities of their tasks deviated from workshop to workshop and from job to job, but examining the chores of a piecer helps illustrate the average workday for an apprentice in a textile factory.

The mechanization of spinning reduced some of the human effort needed to produce yarn, but piecers, who served as human instruments performing the labors beyond a machine's capacity, were solicited in large numbers. For twelve hours a day, a piecer worked in "an aisle several feet wide between two rows of spinning frames," fulfilling monotonous and enervating tasks. As a machine spun and wound thread onto spindles, tension would frequently break the yarn. Whenever a thread snapped, a piecer grabbed the loose ends, twisted them to form a small knot, and snipped the remaining ends with scissors. Those experienced could twist yarns between the index finger and thumb in one swift motion. Although novices required more time for the work, seasoned piecers were able to join dozens of broken threads in a minute. Because the machinery ran without pause, several threads often broke simultaneously, turning up the pressure. The relentless pace of industrial labor and the restriction of personal space, privacy, and freedom of movement were difficult for rural women, born and raised in farming households. Although some adapted easily to factory work, others never did. Kang Pokchŏm conceded that four years of factory life "never made it any easier."[35]


*Life in the Dormitories*

The factory whistle blew in the evenings at a quarter to six as night shift workers came to take over workstations. Another whistle sounded at 6:00 P.M. signaling the start of the night shift. Most spinners, weavers, and silk reelers in colonial Korea were dormitory residents, who, at the end of the workday, returned to housing facilities owned and operated by the company. Dormitories have been seen as intrinsic to female factory labor, but the system was not born with the textile industry or women's wage work. According to E. P. Tsurumi, Japanese executives first considered hostels unnecessary, but in-factory residence facilities proliferated as increasing numbers of women left without fulfilling their contracts. Thus, dormitories were installed in factories that recruited workers in order to reduce absenteeism and turnover. Textile factories in the colonial era, either owned by the Japanese or designed according to their business models, preferred to house their employees.[36]

In early twentieth-century Korea, when much of society still espoused Confucian customs of female confinement, dormitories offered young women and parents the reassurance of safety. Upon entrance, workers quickly discovered that residence halls were not places of protection but corners of isolation. The employees of textile companies could neither leave factory compounds nor receive friends or relatives as visitors. Correspondence was strictly forbidden; the institutions that initially offered parents assurance ultimately thwarted all forms of familial contact.

The dormitory displayed its prisonlike purpose in its construction. It stood adjacent to the mill, within the compound, or fenced alongside it, defended by iron fences eight feet tall, topped with another two feet of barbed wire. The barricade performed two functions: keeping workers in, and contractors, parents, and relatives, who might help girls escape, out. Employees were disciplined inside the halls by residence supervisors (*sagam*), while guards outside kept surveillance for runaways around the clock. Although promises of luxurious living standards might have persuaded young women to start factory labor, the dormitory served as a mechanism for constant control, inside and outside the workshop.

The living conditions offered by most residence halls were spartan. If a woman was fortunate enough to have worked the day shift, she returned to a small room, cramped with twelve to twenty others, where the night worker who shared her few square feet most likely left their living space untidy. At best, two workers shared a six-foot-by-three mat; at worst, over twenty women were squeezed into eight-mat rooms, which measured around twenty-five to thirty square meters. In most cases, two women shared a five-foot by two-foot living area, with a mat, a blanket, and perhaps a small, hard pillow. Each worker was given a set of clothes, one for summer and one for winter, a small block of soap, and, if lucky, a toothbrush. Although dormitories were equipped with communal washing facilities, these were crowded, filthy, and did not allow for thorough bathing. In many cases, as with the Tōyō (Tongyang) Silk-Reeling factory in Chinhae, men and women did not have separate restrooms.[37] Laundry facilities were also lacking, and, for most workers, bathing and laundry were customarily done no more than twice a month, during their biweekly holidays.

While dormitories were first built to secure young women, many factories in colonial Korea came to house all of their employees, men included. Industries that hired unskilled youth from the countryside likewise boarded young men in company residence halls. The Kanegafuchi Spinning and Weaving (Chongyŏn pangjik) mill in Kwangju and the Chōsen Spinning and Weaving (Chōsen bōseki) factory in Pusan, for example, housed their male workers. Although supervision within men's residences might have been more lenient, their housing provisions were just as meager, if not worse. A male participant of the January 1930 Chōsen Textile workers' strike complained: "Thirteen men were cramped into a room made for one." The testimonies of working men suggest that traditional constructions of masculinity did not make the conditions of male boarders any better than those of their female counterparts. During the Chōsen strike, for instance, male dormitory residents complained that unlike the women who received blankets, they were not given sufficient bedding.[38] Stereotypes of masculinity, such as physical strength and stamina, put into practice, sometimes made the working and living situations for men worse.

Restrictions of movement, personal space, and privacy were the norms of factory life for employees of either sex, but only women experienced the distress caused by menstruation. During a time when symbols of feminine sexuality were distanced or altogether ostracized, periods were horrific ordeals for factory girls. Ironically, mills that produced fabric failed to provide women cloth to use for sanitation. Because workers were checked before going to the lavatories as well as to and from the halls, acquiring cotton required invention and cooperation. Kang Pokchŏm, who worked for the Tōyō (Tongyang) factory in Yŏngdŭngp'o, described that groups of women smuggled cotton regularly from storage facilities while friends kept watch. The cotton they obtained was sometimes raw and coarse but better than nothing. Kang explained that if cotton was not used, the dried blood on the uniforms poked and chafed the skin with every move. Still, their physical discomfort probably paled in comparison to the psychological humiliation of being seen with stained uniforms by male workers and supervisors.[39] Although some of the shortcomings factory women confronted were gender specific, other restrictions, such as the ban on leaving factory grounds, were shaped by culturally and regionally specific conventions.

In spite of their distinctions, dormitories in North America, Japan, and Korea were universally likened to prisons by contemporaries and historians. The lives and experiences of the women within have been characterized by powerlessness, misery, and suffering. Nonetheless, the testimonies of former factory women imply that dormitories were sometimes more than prisons. During agreeable times, they were places of comfort and friendship; during difficult times, spaces of commiseration and support. At nighttime, dormitories became "oceans of tears" as girls, discouraged by fatigue and harsh treatment and pained by their memories of home, tried to fall asleep.[40] Despite the impenetrability of their walls, dormitories were also arenas of sympathy, association, and agency. The severity of conditions motivated women to organize and assert their interests; for the greater the struggle, the greater was their need for unified action. Dormitories frequently served as the principal breeding grounds for women's association. Workers' tactics of accommodation and resistance were formed inside, and strikes were often launched in the residence halls themselves.

## Corporate Control, Night Work, Accidents, Disease, and Death

Whether working women were daughters or mothers, in the mill, they were uniformly subject to a system that furthered modern economic development while adhering to age-old paternalistic logic.[41] The factory system maintained a hierarchical order that appropriated the labor and living standards, as well as the rights and liberties, of individuals according to corporate evaluations of skill. A clear distinction was made between skilled and unskilled workers, or manual laborers (*chikkong*). Although women were employed by the company, the people they served on an everyday basis were not the owners but their agents on the shop floor, the supervisors (*kamdokcha*). For workers earning production-based wages, including loom operatives and rubber shoemakers, supervisors assessed individual efficiency and productivity. In theory, daily wages were unaffected by superiors. But the terms of treatment, such as the use of lavatories or the taking of breaks, were mediated by the permission of supervisors. Overseers frequently treated their workers on subjective bases. Kang Pokchŏm recalled that better-looking women were granted more breaks and were subject to fewer beatings.[42] On the other hand, she mentioned, attractive workers were at greater risk of sexual harassment and rape.

As the next chapter will explore, many historians have assumed that labor disputes in the colonial era were provoked by ethnic and cultural tensions. Nonetheless, even a cursory review of strike demands shows that workers did not protest against Japanese management in general but appealed for the dismissal of specific wrongdoers. One such example was the Kunze Silk Reeling (Kunsi chesa) strike involving 650 male and female workers in Taejŏn, which broke out on November 7, 1932. Calling for the dismissal of the main superintendent and the improvement of meals and wages among other proposals, the strike lasted for eight days, resulting in the firing of the superintendent.[43]

Just as overseers on the shop floor enforced the company's rules, so did dormitory supervisors. Whereas overseers were men, most dormitory supervisors in textile enterprises were women a few years older than the workers. Unlike foremen, who exercised authority by force, these women wielded moral influence. They were most likely avoided rather than feared, as dormitory supervisors were the first to report secret meetings or other signs of resistance among the boarders. Although residence advisors were caretakers, they were also agents of the company. Grievances against them incited boycotts. The 1931 strike of the workers of the Kunze Silk Reeling factory in Ch'ŏngju, for instance, demanded the dismissal of a dormitory supervisor who alerted the managers of the residents' clandestine plans for revolt.[44]

The factory system placed women workers in an unprecedented social context, squalid living conditions, and unyielding labor, never before experienced by most peasant girls. The logic of mechanized manufacturing required women to work in fixed shifts and to keep "pace with machines, in an environment where the temperature and speed of work adjusted to the needs of cotton thread rather than human beings."[45] Instruments that were intended to maximize output became their guides. Twenty-four hours a day, seven days a week, carding and drawing machines, ring frames, looms, spindles, and bobbins ran in perpetual motion. Every four or five years, mechanical failures induced by such strain caused fires. But executives defended their practices, arguing that even if factories burned down, the cost of shutting down the machinery daily was still greater. To accommodate the machines running continuously, women worked at night.

By the 1930s, the night shift for women and children was abolished in Japan, but these rules did not apply to Japanese-owned factories in China, Korea, Manchuria, and Taiwan.[46] In spite of strong opposition from Korean intellectuals and Japanese reformers, in colonial Korea, night duty (*yagŭn*) for women and children prevailed. Some employees preferred working at night because of the slower pace, but mandatory evening shifts interrupted young sleeping schedules. Lack of sleep, however, was merely one of the many side effects of night work. Contemporary researchers such as Ishihara Osamu found that women underwent erratic cycles of weight loss and gain as they shifted from night to day shifts. Prolonged periods of night work made youth more vulnerable to contagious diseases such as tuberculosis (*p'yegyŏrhaek*). Although night workers were given the same meals as day workers, they not only lost more weight but were also more disposed to malnutrition (*yŏngyang pulyang*) and diseases caused by dietary deficiency.

For many, the physical demands and dangers of night work were overwhelming. In spinning mills, carelessness often cost "a tired girl a finger."[47] But, as increasing numbers of women were employed in all divisions of textiles including the most dangerous, workers not only lost fingers and limbs but died in the factory. Night work exposed women to sexual harassment, which was possible by day but more frequent at night. Many workers claimed that bodily injury was preferable to the humiliation of rape. Because such incidents caused much controversy, both parties generally concealed them. After such assaults, young women were undoubtedly consumed with fears of pregnancy and of the obstacles they might encounter in finding future husbands and, of course, with unimaginable shame. Former workers testified that pregnancies were "taken care of" by the company, but at the worst extreme, women administered their own abortions or took their own lives. Yi Chaeyun, who worked at the Kanegafuchi factory in Kwangju recalled that one night, she and other workers saw a girl being taken into a supervisor's bedroom. The next day, they learned that she committed suicide thereafter by "hanging herself with tape."[48]

Because many textile factories in colonial Korea were Japanese owned and run, the recruiters and supervisors that mistreated young women were most likely Japanese. This fact has been underscored by popular historians to blame Japanese colonization for corporate abuses. But accidents, disease, sexual abuse, and death were not simply provoked by ethnic tensions. The dormitory system installed in Japan and Korea might have increased work-ers' vulnerability, but women in other regions were similarly at risk of as-sault on the shop floor, in their homes provided by contractors, and on their way to and from the job. Expanding the scope of analysis beyond the colo-nial experience shows that gender-specific abuses in industrial labor were not necessarily imposed by the Japanese as accidents, disease, and sexual abuse continued in Korea after Liberation.

## Summation

Across cultures, industrialization ushered in unparalleled opportunities and unimagined exigencies for young rural women. In early twentieth-century Korea, the usual rigors of modernization were compounded by a resilient tradition of patriarchy, colonial governance, and accelerated capitalist de-velopment. Without doubt, Korean factory women's struggles were inspired in part by colonial structures and capitalist strategies. Contemporary labor activists and intellectuals have argued persuasively that wage labor and, in particular, mechanized labor under colonial capitalism subverted human es-teem and individual liberty. Nevertheless, arguments that fault colonial cap-italism unilaterally only detail the story in part, aggrandizing hegemony from above. For sure, the experiences of Korean women were mediated by the political contexts surrounding them, but questions of alternative influ-ences remain: Were these workers also manipulated by their personal ties? Did familial obligations also undermine the human rights of autonomy and dignity? In short, why did factory girls endure such herculean labor and liv-ing conditions?

A critical analysis of colonial factory women, their endurance in the mills, and their resistance against their employers indicates that the lives and labors of factory women were shaped by two overarching forces: the politi-cal and the personal. The rapid growth of the manufacturing sector after WWI brought light industries, such as silk reeling, cotton spinning, weav-ing, and knitting, as well as food and rubber processing, to the fore of the economy. Although these businesses hired women, their recruitment, management, and productive methods as well as the makeup of their work-forces varied by industry, giving rise to distinct types of female factory labor. Married women dominated the urban rubber and knitting sectors, whereas unmarried women, often girls, were systematically recruited for employment in large spinning and weaving firms. Certainly, macro political structures and historical progressions situated women within strict economic and social confines. Nonetheless, a survey of the conditions of recruitment, contracts and wages, work inside the factory, life inside the dormitories, night work, and calamities suggests that women's lives and labors were determined by interpersonal networks such as those formed on the shop floors and within the dormitories. Their routines inside the workshops illustrate how women endured industrial labor, whereas their conversations in the dormitories shed light on why: women's dreams paralleled their waking ambitions as they toiled to support their families.

Although research on early twentieth-century Korean workers has tended to converge on large-scale demonstrations, women's agency was not limited to conflict. Resistance was only made possible by endurance. Searching for the forces that sustained the perseverance of factory women evokes queries concerning the dynamics between the political, the personal, and the individual. Was consensus manufactured by colonial capitalism, or was consent the product of traditions and values closer to the individual? These questions of compliance are further explored in the next chapter examining the antithesis of endurance—resistance.

FOUR

## Contests of Power and Workers' Modes of Association

Capitalist industry flourished throughout the colonial era, but so did labor conflict. As soon as production began, factories were shut down by riots, boycotts, sit-ins, and other forms of contention. Should readers ever thumb through the headlines of popular newspapers such as *Tonga ilbo* (East Asia Daily), *Chosŏn ilbo* (Korea Daily), and *Keijō nippō* (Kyŏngsŏng Daily) during the interwar years, they would be struck by the sheer frequency of labor strikes.[1] Whereas in 1921, 3,403 workers incited thirty-six strikes, by 1931, 201 strikes involving 17,114 workers erupted. In the mere span of ten years, workers' activism emerged as a formidable challenge to capitalist development under colonial rule.

Of the labor demonstrations of the interwar period, few have received as much attention as the Wŏnsan general strike of 1929, involving over 2,000 workers. Scholars such as Hagen Koo characterize the event as a "climax" of workers' unrest, whereas others including Soon-Won Park mark it a "beginning." The Wŏnsan strike served as a turning point in the history of colonial labor activism, catalyzing unprecedented collective action in the Depression years that followed. It also signified the height of communist-labor infiltration in the north, particularly in the South Hamgyŏng and South P'yŏngan provinces. Park maintains that through communist leadership, workers' petitions "became more sophisticated, and their solidarity and willingness to hold their positions" not only multiplied their strikes but also magnified their impact. Although the Wŏnsan strike displayed the organizational capacities of Korean workers, its legacy was more far-reaching than its immediate effects.[2]

At the outset of WWI, much of Western Europe, North America, and Japan had generations-long histories of industrial labor, thus the interwar demonstrations surfacing in these regions have been perceived as expressions of proletarian consciousness. If, however, workers in colonies protested against capital, how should historians interpret their consciousness? Would their identities be anti-imperialist or nationalist before they were proletarian? The intensification of labor activism in Korea offers an opportunity to examine the complexities of workers' identities and alliances in a late-developing, colonial context. The temptation to concede to the unity of class or nation diminishes, however, when the analytic, hermeneutic, and epistemological problems inherent in conceptualizations of class or nationalist consciousness are unraveled. Does the proletarian paradigm, founded on European liberalism, and concepts of autonomous states and citizenry, explain the consciousness of workers in colonial Korea? Does nationalism, engendered by urban reformist elites, represent the motivations of factory workers? What happens when age- and gender-specific disparities are explored in conjunction with such assertions of collectivity?

Whether these protests evidence the formation of a modern working class (*nodongja* or *kŭlloja kyegŭp*) or whether popular working-class consciousness (*nodongja kyegŭp ŭisik*) inspired these protests remains contested. Social scientists and specialists of late twentieth-century labor movements assert that though select activists sought to foment working-class consciousness in response to Japanese imperialism, class solidarity was not widespread until after Liberation in 1945. Likewise, the more conservative historians of colonial Korea maintain that early twentieth-century labor activism was largely inspired by nationalism. Those further to the left of the political spectrum, including North Korean scholars, argue the converse. According to their view, because nationalist and labor activities were harshly suppressed from the late 1930s, Koreans were unable to fully express their proletarian consciousness until after Liberation. Marxist historians claim that colonial workers held a consciousness distinct from those of elite nationalists, which gave birth to the rise of communism in North Korea. Although scholars are divided over the preeminence of class or nation, historians of labor in colonial Korea generally agree that during the first half of the twentieth century, industrial capitalism and a proletarian class emerged concurrently.

In early twentieth-century Korea, the proportions of female and child workers in factories were markedly high. In 1925, after less than a decade of modernization, women composed 24.8 percent of the labor pool in medium- to large-scale factories. By 1930, this figure rose to 33.7 percent— a stark contrast to the composition of women in the mechanized workforce in the United States, which was a mere 13 percent in 1930.[3] Because socialist and nationalist historiography has tended to minimize or ignore the boundaries between and within groups, the political activities of Korean women and children have been seen predominantly as contributions to, or components of, larger, masculine, nationalist, and proletarian initiatives.

This chapter offers an alternate and intimate interpretation of factory women in colonial Korea and their avenues of action and empowerment. First, it examines some of the largest general strikes of the early colonial era (1910–37) organized by women in medium- to large-scale enterprises such as rubber shoemaking, cotton spinning, and weaving. Second, it analyzes protests in the silk-reeling and knitting sectors, unearthing the varieties of workers' defiance and the intricacies of their alliances and identities. Although the Depression stimulated a sharp surge of labor demonstrations, how workers coalesced and presented their proposals diverged according to, among other factors, industrial-, local-, and labor-specific conditions. Rubber shoemakers, for example, gathered in large-scale, citywide, general strikes, whereas the spinners and weavers of cotton textile plants carried out factory-wide boycotts. The demonstrations of workers in both sectors were among the largest, best organized, and most widely documented uprisings of the colonial era. Mill operatives in knitting and silk-reeling projects, lacking the massive labor pools of rubber and cotton factories, protested more creatively. While knitting plants were small, they were decidedly urban.

Thus, hosiery workers rallied in interfactory, citywide strikes. Silk reelers, fewer in their numbers and situated further apart from each other, instead, organized interfactory, company-wide boycotts. Striking in rounds throughout the various mills of the same company strengthened the bargaining power of silk reelers.

The collective activities and assertions of these workers clearly demonstrate that Korean factory women were not mere handmaidens of the masculine, colonial labor movement. Just as significantly, they unveil the extent to which both male and female labor activism was spurred by contextually changing assessments of skill and value. Cross-referencing the definitions of skill as understood by contemporaries, historians, and historical actors reveals that women's identities were not marked by proletarian but by what I term *labor consciousness*, that is, identification with the various forms of paid and unpaid duties performed in the constantly shifting contexts of the life course. In spite of the fact that workers' activities have been associated with colonialism, capitalism, and patriarchy, it was communities, not structures, that propelled protests. These communities were, as Victor Turner describes, often "unstructured or rudimentarily structured, [with] a relatively undifferentiated comitatus." Rather than systematic, the diversity and specificity of their designs point to the possibility of conceiving women's labor activism in colonial Korea as acts of "liminality."[4]

United as a factory, workers called for improved wages, conditions, and treatment. Seemingly isolated strikes often prompted industry- or city-wide boycotts that halted production for months at a time. But did women embrace the ideals of the men who comprised over two-thirds of the total labor force? Not altogether. Universal notions of suffrage for women or workers did not necessarily incite their boycotts. Rather, the labor consciousness of Korean factory women relied on their internalizations of knowledge and skill as well as their instruments of power, that is, their modes of association inside the mills. Arising from "locality- and work-unit-based" organizations of interests as well as the intra-industry and intrafactory divisions that gave way to interest groups, factory women in colonial Korea engaged in what Ching Kwan Lee calls "cellular activism."[5] Whether as a pair, a group, a factory, or a workforce, collaboration enfranchised women and allowed them to disrupt normal power relations that could have resulted in their benefit. On the shop floor and in the dormitories, women fostered friendships and alliances that regulated the conduct of superiors and each other. Finally, this chapter posits that factory women's modes of association assisted their compliance and resistance in the mills as well as their departure. Reviewing their rationales for and methods of escape confirms that together, women not only negotiated their terms and conditions, but just as important, eluded these terms altogether.

## General Strikes of Women in Rubber and in Cotton Spinning and Weaving

That contemporary intellectuals underscored the significance of general strikes (*ch'ongp'aŏp*) such as Wŏnsan and other male labor-specific boycotts led historians to conclude that workers' political power was connected to masculinity and patriarchal standards of skill. The protests of female factory workers, nevertheless, show that the organizational procedures of laborers were determined more by culturally, historically, locally, and industrially specific contingencies than by their class, ethnicity, or sex. Rubber processing and spinning-and-weaving mills, for example, dependent on constant access to imported raw materials, were generally based in urban areas. The particularities of manufacturing mattered for spinners, weavers, and shoemakers who were able to achieve large-scale, centralized organization.[6]

Like their male counterparts, factory women formed unions from the outset of Korean modernization. Established on July 3, 1923 by more than 200 members, the Kyŏngsŏng League of Female Rubber Workers (Kyŏngsŏng komu yŏja chikkong chohap) was one of the first exclusively female labor unions on the peninsula.[7] While some organizations were initiated for offensive reasons, others were defensive. The Pyongyang League of Young Women Hosiery Workers (P'yŏngyang ch'ŏnyŏ yangmal chikkong chohap), for instance, was founded on April 11, 1925 in response to shop floor problems and pay cuts at the Pyongyang Hosiery factory (P'yŏngyang yangmal kongjang).[8] Among others, the Wŏnsan Association for Female Friends of Labor (Wŏnsan yŏja nouhoe) was established in July 1924, and the Pusan League of Women Workers (Pusan yŏgong chohap) was inaugurated in April 1929. Although they exemplified the vitality of women's labor organization in the 1920s, few of these leagues remained active after the early 1930s.

Collectivism enabled rubber shoemakers perhaps more than the employees of other female labor-intensive projects. Two of the largest labor demonstrations of the colonial era, the general strikes of rubber workers in 1930 Pyongyang and 1933 Pusan, convey the extent of women's activism in the chemical industry. The first strike, opposing pay cuts resulting from market recessions, erupted in a demonstration involving the employees of several rubber factories in and around Pyongyang in August 1930.[9] The protests disrupted production for twenty-three days and involved over 1,800 workers from nine different companies. Much like the 1930 strike in Pyongyang, the issue of wages ignited another general strike in 1933, this time in Pusan. Responding to the Mitsui takeover of numerous rubber mills in South Kyŏngsang province, on October 17, 1933, 130 workers of the Taehwa rubber factory walked out in opposition to the company's retraction of the former wage scales. The next morning, 280 female workers of Ilyŏng Rubbers in Ch'wach'ŏndong and 120 workers of Nŭngam Rubbers followed. By October 20, the employees of Pusan Rubbers, Hŭisŏng Rubbers, and Yulchŏn Rubbers joined in the unified boycott protesting the pay cuts imposed by the Mitsui conglomerate. The Pusan strike of 1933 resulted in a somewhat favorable compromise for the workers, whose daily wages were docked by 0.04 rather than 0.045 yen. Although the dispute ended in mutual concession, the Pusan rubber workers' general strike showcased the organizational capacity of women in concentrated urban settings.[10]

These demonstrators' demands articulated not only designs for immediate relief but also long-ranging plans for the future. As the 1930 Pyongyang workers' list of grievances illustrates, though wages were the primary concern, far-reaching aims such as extended holidays and "financial compensation for work-related injuries" were also voiced (Figure 4.1). Women proposed gender-specific provisions for "three weeks of leave before and after childbirth (ch'ulsan)" and "unrestricted time for breastfeeding" on the job. The strikes of the 1930s proliferated in quantity and matured in quality as workers appealed for universal objectives including the eight-hour workday and disability compensation. That, in 1930, female factory operatives called for the revision of abuses that continued after Liberation, including

*Figure 4.1* Demands of the Pyongyang Amalgamated Rubber Worker's Union
(P'yŏngyang komu chikkong chohap) (1930)

1. Reversal of wage cuts
2. Reinstitution of those fired without reason
3. Improvement of the supervisory treatment of workers
4. The institution of paid rest days and holidays
5. Abolition of night work
6. Reduction of exploitative work hours
7. Equal distribution of shoemaking materials
8. Compensation for work-related injuries
9. Abolition of the system of making workers responsible for machinery repairs
10. Termination of irregular/unnoticed inspections
11. Abolishment of disciplinary measures and the penalty wage system
12. Abolition of the system of workers having to pay for inferior products
13. Abolition of the compulsory savings system
14. The institution of free tools of the highest quality
15. Abolition of cleaning jobs for workers
16. The provision of year-end bonuses
17. Three weeks of holidays before and after childbirth (*ch'ulsan*)
18. Freedom or unrestricted time for breastfeeding
19. The settlement of collective recommendations and collective agreements

Source: *Tonga ilbo*, September 10, 1930.*
 * Some 1,800 workers from nine companies sent their demands to the heads of administration for adoption. *Tonga ilbo*, September 10, 1930.

night work and production-based wages, captures the prescience of working women in colonial Korea.

While the employees of rubber shoemaking projects, working in close settings, were able to assemble in citywide general strikes, cotton spinners and weavers more frequently organized in industry- and factory-specific manners. One of the largest demonstrations of the colonial era was the 1930 general strike among the employees of the Chōsen Spinning and Weaving (Chōsen boseki or Chōbō) factory in Pusan. The boycott lasted almost a month and involved more than 2,270 workers. The demands presented to the heads of Chōbō, then the largest textile plant in Korea, relay the greater objectives of women workers. (See Figure 4.2.) Chōbō em-

*Figure 4.2* Demands Presented to the Owners of the Chosŏn Spinning and Weaving Company (Chōsen boseki kabushiki kaisha) (1930)

1. Increase in wages
   a. The establishment of a system of providing a minimum of 0.80 yen for daily wage laborers (*ilgŭp kongsu*)
   b. For the wages for contracted workers to be increased by 3 percent
   c. For the raise of 0.80 yen to be instituted and established as the official daily wage (*hyŏnilgŭp*)
   d. To improve the mill's working conditions to match those of factories in Japan and for the institution of paid rest hours
2. Institution of the eight-hour workday
3. Establishment of a promotion system
   a. Institution of a system which provides workers with 0.05 yen raises after four months of labor
   b. To establish the lowest pay raise as 0.05 yen or higher
4. The elimination of the current system of dismissal
   a. Complete termination of the system of dismissal without notice (*mudan haego*)
   b. Institution of a system of providing a year's salary for workers dismissed by force (*pudŭgi haego*)
   c. For the company to notify, one month in advance, those to be dismissed forcefully
5. For work tools to be given free of charge
6. Compensation to be given to those suffering from misfortune, e.g., those fired or those grieving the death of relatives
   a. For the company to provide three years' wages for persons injured (*pulhaengja*) on the job
   b. For the company to provide a year's living costs for the families of those who died (*samangja*) on the job
7. Elimination of the penalty system
8. The improvement of meals
   a. Provision of second-grade rice
   b. Provision of fish or meats during lunchtime
9. Payment for private housing
10. Elimination of discrimination or differences in treatment of Japanese and Korean workers
11. Elimination of the current regulations on entering and leaving premises for nondormitory residents
12. The fulfillment of these demands
    a. To refrain from dismissing any worker because of this incident
    b. In the case that a worker is dismissed because of this incident, to provide 100 yen in compensation

Source: *Tonga ilbo*, January 17, 1930.*
\* Although working women called for treatment equal to that of Japanese workers (Demand 10), demonstrators presented their proposals according to their priority. Therefore, workers more likely sought the former demands, such as the proposal for working conditions to match those of factories in Japan (Demand 1d), but did not expect the latter demands (Demands 10, 11, and 12) to be fulfilled.

*Plate 4.1*   Weaving room (1930s).

ployees called for the institution of eight-hour workdays and proposed improved shop floor standards. Although they were concerned with the company's rules about their tools and equipment, working women's sights were also fixed on wages. Demonstrators called for the establishment of a minimum wage of 0.80 yen a day. Interestingly, their target was 0.16 yen lower than the average daily wage for Japanese women workers in 1930 Korea, suggesting that Koreans did not expect to be paid as much as the Japanese but sought fair compensation. In addition to minimum wages, women wanted their earnings to be protected from inflation rates that escalated dramatically in the years following the 1929 crash. Wages sparked the strike, but ultimately, workers requested that daily salaries be raised only 0.05 yen.

The textile industry, which relied heavily on unskilled female workers, was characterized by the high turnover rates of its employees. Therefore, an overriding concern for women in Korea was job security. While work in spinning and weaving factories was "easy to get and easy to lose," women were not complacent about these insecurities. Especially despised practices were dismissals without notice (*mudan haego*) and forced dismissals (*pudŭgi haego*). Thus, proposals such as compensation for injured workers (*pulhaengja*) and for the families of those who died on the job (*samangja*) sought assistance for accidental costs.[11] Although the Chōsen Spinning and Weaving Company was Japanese owned, workers' appeals focused neither on the preferential treatment of Japanese employees nor on equal wages. Working women did not necessarily strike because of imperial rule. Rather, their protests vocalized some of the pressing problems for factory workers during the early phases of mechanized development in colonial Korea.

## Strategies of Women in Knitting and Silk Reeling

Like rubber operatives, the employees of knitting factories, particularly those in Pyongyang, Pusan, and Seoul, were better able to collectivize regionally. The Pyongyang hosiery workers' strike of 1932, which involved the majority of the knitting factories in and around the city, exemplifies the efficacy of these women's organization. Between January and March 1932, fifteen rounds of boycotts broke out, crippling the Pyongyang knitting industry. In response to booming hosiery sales in Manchuria, knitting operatives petitioned for pay raises to be effected in thirteen factories, including the Chosŏn, Samgong, Taedong, Sech'ang, Sinwŏn, P'yŏngyang, and Ŭich'ang companies. (See Figure 4.3.) Whereas volatile market fluctuations incited the 1930 rubber workers' strike, heightened consumer demand for knitted goods inspired the 1932

*Figure 4.3*  Demands of the Workers of the Pyongyang Taedong Hosiery Factory (P'yŏngyang Taedong yangmal kongjang) (1932)

1. A 3 percent increase in wages
2. Provision of the cost of needles
3. Elimination of the penalty system
4. Reduction of work hours
5. The completion of the dining hall and the establishment of separate toilet facilities for male and female workers
6. The abolition of night shifts, and if forced, the introduction of a 3 percent pay raise for night labor
7. The elimination of the system of dismissing employees in groups
8. Provision of slippers for dormitory workers
9. The adoption of these demands

Source: *Chungang ilbo*, March 14, 1932.

rounds of hosiery workers' protests in Pyongyang. In the end, several of the factories increased their employees' wages; thus, the hosiery workers' strikes of 1932 verify that women workers were able to exercise offensive power.[12] Although the demonstrations of March 1932 were curtailed, another round of boycotts was launched in June, as the managers of Taedong, Siwŏn, Sech'ang, and Ŭich'ang hosiery factories retracted their earlier raises of 0.02 yen. The imperial police ended the demonstrations of June 1932, but in the end, Sech'ang Hosiery workers obtained a raise of 0.05 yen in their daily wages.[13]

In terms of their work hours, pay, and supervision, silk reelers were perhaps the most dispossessed of the factory workers in colonial Korea. By the early 1930s, most cotton textile factories in Korea operated in twelve-hour shifts. But because silk production did not rely on continuous machinery, reelers frequently worked up to thirteen and fourteen hours a day. Throughout the latter half of 1932, a wave of demonstrations erupted in filatures throughout Korea, and the most urgent concern of these workers was the reduction of their shifts. The employees of the Katakura silk-reeling factory in Yŏnghŭng boycotted work because they could no longer endure fourteen-hour days, "from four in the morning to six in the evening." Despite the degrees of their disenfranchisement, silk reelers contested the authority of owners and managers. A *Tonga ilbo* article detailed: On August 20, 1932, over 500 female employees of the Chosŏn Silk Reeling mill in Seoul walked out in opposition to wage cuts, long hours, and the forced dismissal of two workers.[14] The company reinstated the fired women but refused the majority of their proposals. Thus, on August 30, workers halted production again, calling for the reduction of the thirteen-hour day, the introduction of daily wages, and better treatment. Although the second strike ended in defeat, it evidenced that these women, after initial success, sought to improve their odds.

Silk-reeling mills, dependent on silkworm and mulberry farms, were located in more remote cities such as Ch'ŏngju, Taejŏn, and Yesan; therefore, women employed in the silk industry did not always enjoy settings favorable for labor solidarity.[15] Dispersion, however, did not mean isolation. Because they were detached, silk reelers found it more effective to protest circuitously, or in rounds. News of an uprising in one factory often prompted workers in another to strike. In late November 1932, for instance, inspired

*Plate 4.2*   Silk-reeling workers (1930s).

by the success of the Michize silk reelers' boycott of the previous week, the women of Kanegafuchi (Chongyŏn) silk-reeling factory in Kwangju struck. As chronicled in a newspaper report, the Kanegafuchi workers' demonstration was a "shadow and echo" of the earlier strike at Michize (Tosi) Silk Reeling. Whereas proximity assisted the activities of the Kanegafuchi and the Michize silk reelers in Kwangju, the strikes of the workers of Katakura Silk Reeling in Hamhŭng and Chŏnju illustrate the effectiveness of inter-factory collaboration. On July 30, 1932, at 5 A.M., 420 female workers of the Katakura silk-reeling factory in Hamhŭng refused to start the workday. They called for reduced work hours, more nutritious meals, reformed work environments, and proper treatment. By the next day, the company detained the women in the dormitories with police assistance. Over thirty women escaped captivity but were quickly found by the authorities and returned to the factory. As a result of their perseverance, many of their appeals, such as reduction of work hours and the reinstatement of workers dismissed for running away, were met. Using the Hamhǔng demonstration to their advantage, women workers of the Katakura factory in Ch'ǒngju struck offensively for raises in August 1932. Although their acquisitions were temporary, the employees of the Ch'ǒngju factory won the round by using both conventional and innovative tactics of resistance. The strategies of women in knitting and silk-reeling projects convey that the scale of a strike did not signify its success.[16]

Although many studies on colonial Korean labor have emphasized general strikes in heavy industries, workers' abilities to organize large-scale demonstrations were shaped by the characteristics of the industries that employed them. Businesses requiring access to natural resources, such as mining and oil refining, established their bases in northern cities generally close to reserves. Those reliant on imported resources, including hosiery and rubber shoe manufacturing, set cornerstones in port cities throughout the peninsula. Because most rubber shoemaking factories were founded in large urban centers, particularly in and around Pyongyang, Pusan, and Seoul, their employees were better able to collectivize and integrate workers in separate factories through unions. The colonial knitting industry likewise formed its bases in these cities, unwittingly assisting its workers' modes of resistance. Whereas women in rubber and hosiery manufacturing achieved greater solidarity through their close proximity, the employees of spinning, weaving, and silk-reeling enterprises frequently did not enjoy such favorable settings. Large cotton textile plants were built in cities, but smaller spinning mills were dispersed throughout Korea. Filatures were even more remote. Therefore, workers in smaller cotton-spinning and weaving and silk-reeling factories were often less able to coordinate general strikes.

The majority of labor demonstrations in colonial Korea were small, factory-specific events.[17] Interfactory and interindustry collaboration facilitated seemingly isolated boycotts, and the details of these demonstrations show that workers' organizational capabilities did not always take "general strike" forms. Workers in certain sectors, such as those in knitting and silk reeling, found it more effective to strike in rounds, whereas rubber shoemakers secured greater efficacy through general mobilization. Although working women protested against gender-specific abuses, specific corporate and shop floor climates also configured their methods of association and assertion. The activities of women in rubber shoemaking, silk reeling, and cotton textile manufacturing display not only the formative stages of female labor activism in Korea but also the circumstantial ways in which most strikes emerged in the colonial era.

While the labor protests of Korean women in the early twentieth century are frequently depicted as defensive acts, a close reading of their petitions indicates that they often struck offensively, to improve their work and living conditions, avenues of association, and bargaining power. The most successful offensive strikes were those directed against individuals. The 1930 strike of the employees of the Katakura silk-reeling factory in Seoul, for example, resulted in the quick dismissal of an unfit supervisor. Impediments to their association also motivated workers to walk out. The 1931 strike by the employees of the Pusan Marufuto (Hwant'ae) rubber factory was incited by the company's forced termination of a savings and loan association (*kye*) among its workers. Women also called for the establishment and expansion of social facilities. The workers of Ch'ŏngju Kunze (Kunsi) silk-reeling factory, in their three-month-long boycott, appealed for the building of an athletic field.[18] Just as they resisted pay cuts, women steadfastly protected the avenues of association that enabled them. Faced with the opposition of the company and police, workers pulled together, disregarding their differences. In daily life, however, women associated with each other in smaller groups.

Consciousness-raising campaigns, nationalist and socialist, influenced factory laborers in part, but their effects were not lasting. Although in times of economic depression, political ideologies fused with workers' real demands, most factory women did not obtain much assistance from labor unions or leftist infiltration. Usually bound to company grounds, women did not meet with union leaders but befriended each other in workshops and dormitories. The vast majority of labor demonstrations incited by women in colonial Korea rose spontaneously and independently. In the face of greater opposition, workers united and called for solutions to the specific problems of their labor. More often, however, women manipulated or resisted authority in less conspicuous ways.

Whether they struck or not, working women in colonial Korea were influential figures inasmuch as they exercised some power over their lives. Examining how identities were formed by the actions and labors internalizing skill reveals the specificity of workers' consciousness. Individual definitions of skill were determined by not only broader public requisites, such as standards of training and experience, but also personal, "task-specific" memories that affiliated individuals to groups of workers. Recognizing that skill required internalized ideas and activities prompts reconceptualizations of the workings of power from below. Just as women acquired benefits through protest, they exercised agency by forming and reforming the systems of inclusion and exclusion inside the mills, using traditional avenues of association, such as similarities of background, age, and areas of origin.

## Acquisitions of Skill and Practices of Power

For employers in colonial Korea and elsewhere, skill was a quantifiable commodity. Apart from ability, an individual's age, ethnicity, experience, height, level of education, overall health, sex, social background, and weight were some of the factors determining a worker's value. In the 1920s and early 1930s, mills simply divided their employees into skilled and unskilled groups. By the late 1930s, however, many factories adopted elaborate, hierarchic gradations of classifying their staff. Novices, apprentices, and other unskilled employees, who were usually Korean women and children, were at the lowest levels of company hierarchies. Skilled, ordinary (*pot'ong*; *futsū*), or "regular" workers were subdivided into three categories of "second, first, and superior." Classed above ordinary workers were overseers, dormitory supervisors, and shop floor managers (*yŏkpu kongin*; *yakuzuki kōin*). Factory women, especially those first recruited as children, usually began as novices performing simple tasks such as piecing thread in textile factories, earning the lowest wages.[19]

For employees, however, skill was qualitatively constructed. Proficiency was assumed rather than articulated; and this assumption was reinforced by ritual, language, and oral tradition. Participation in ceremonies with fellow craftspersons and attitudes toward outsiders or strangers defined the boundaries of workers' alliances. Codes of dress, conventions of conduct, and familiarity with the instruments of their craft symbolized proficiency. Although often without formal certificates of skill, working women, like craftsmen, claimed mastery by adopting symbols and rituals, such as special forms of clothing and tools, which displayed expertise. The demonstrations of women workers in colonial Korea affirm that they not only assumed but articulated their proficiency as they frequently appealed for workshop tools of the highest quality. For men, adeptness might have been inherited property, but for women, skill was accumulated capital.[20]

The language employed in their testimonies of skill and their strike demands articulates the political consciousness of working women in colonial Korea. As linguistic turn theorists have argued, language, as a system of signs, or symbols of meaning, provides heuristic windows into envisioning how societies conceptualized their realities. Although class and nationalist consciousness was imposed on them, women of the colonial era did not articulate their struggles as associated with class (*kyegŭp*; *kaikyū*). Repeated in their testimonies were, rather, references to *nodong* (*rōdō*), meaning "labor" or "effort." As their expressed identification with "diligence" (*kŭllo*; *kinrō*) suggests, for female factory operatives, "labor" meant "industriousness." For most wage laborers in early twentieth-century Korea, work did not translate into participation in proletarian campaigns but signified toil and exertion. Inasmuch as they identified with the tasks they performed most regularly, the identities of working women were shaped more by their labors than by their larger economic or ethnic affiliations.[21]

At the fore of women's labor consciousness were internal assessments of capacity and productivity. Three broad classes of abilities have been intrinsic to skill acquisition: general intelligence, perceptual speed, and psychomotor ability. Although intelligence might have been an important factor in first learning a task, the task was mastered through repetition, which quickened the timing of perception and reaction; thus, it was through repetition and practice that skill was refined.[22] Proficiency was also an outgrowth of the culture that first prompted the individual to work. Thus, markers of merit were rarely fixed. Employees, by adapting to the routines of the factory or the workshop and observing its customs, could increase their value by hewing to the company's ideals of skill. Collectively or individually, they could sustain and even raise their income and job security. The norms adopted by employees in turn helped configure managers' and owners' standards of expertise. Although owners set the benchmarks of proficiency, if they were too high, workers protested and reset the bars. Still, strikes were only extensions of negotiations on the factory floors. As they contested official constructions of expertise and value in times of dispute, in times of seeming calm, workers accommodated, mediated, and resisted authority in smaller groups. Standards of skill, rates of efficiency, and indeed, balances of power inside the dormitories, workshops, and factories were never fixed. Labor unrest was generally followed by an uneasy peace, which eventually faltered, triggering the cycle of resistance again. Amid such uncertain climates, women formed systems of inclusion and exclusion inside the mills that helped them impress owners, supervisors, and each other. In the face of greater opposition, they united and protested for improvements specific to the conditions of their labor. In everyday life, they exercised power through their daily negotiations within the webs of influence inside the factory.

Taking the liberal self as their point of departure, many scholars remain affixed to the primacy of hegemony by concentrating on limits to individual autonomy. Whether individual or institutional, however, power is always confined by circumstance. That choices are restrained, whether for individuals or groups, does not mean that those with limited choices are more reactive than active. Moreover, all forms of dependence offer resources whereby those subordinate can affect the activities of their superiors.[23] Because individual "selves" within social settings are, in fact, hierarchies of identities that transform over time, circumstance, stages in the life course, and social context, people might acquiesce to the detriment of less conducive identities while struggling to defend the integrity of more important ones. Accommodation can thus be seen as an "investment in subordination"—a situation in which a person might resist taking on one role for the future security of another. Late twentieth-century psychological and sociological research, which found that people prioritize their identities and behave accordingly, strongly suggests that historical actors were more than passive recipients of knowledge and power. Despite their limitations and failures, the adaptive abilities of ordinary people indicate that individuals exercised agency by realizing their most salient goals.[24]

That workers gained leverage through unified demonstration is well known. In daily life, workers' associations of interest and arenas of power were smaller. Groups of women contested factory discipline, but they also competed with each other. Conflicts were common among female operatives: urban factory workers vehemently opposed the hiring of rural female recruits; housed workers envied day workers; women of varying skill and rank stayed away from each other; women of similar backgrounds kept to their own; women of different ages remained apart. The protests of working women generated the attention of contemporaries and historians, but their acts of resistance were not always united. Organized strikes were not the sole barometers of workers' collective authority. More pervasive were familiar associations and alliances within factories and dormitories, which provided the means for resistance. Although women cooperated and demonstrated collectively against mistreatment, more often they endured, challenged, and defied obstacles in less radical manners. Employees not only adapted to company mandates through companionship but also resolved the problems of factory life together. As owners and managers devised methods of discipline, factory women likewise formed strategies of struggle. Further examination of two instruments of workers' agency—their modes of association and running away—expose some of the ways in which Korean women exercised power in everyday life.

## Agency Through Association

Economic development in twentieth-century Korea took place in a shorter span of time than in earlier modernizing nations. Thus, the colonial Korean worker was not born into an employee's but a farmer's family. As taxes, the loss of land, and tenancy escalated throughout the 1920s, increasing numbers of impoverished peasants migrated to urban centers for wage work to supplement the household economy. In most cases, young men and women obtained employment in factories for three to five years, contributing to household incomes and saving funds for marriage. Industrial workers in colonial Korea, therefore, most likely did not perceive the city as a place of permanent residence. Rather, their urban sojourns were temporary as they looked toward the future while biding time in the present. For those confined to factory premises, feelings of displacement and imprisonment were often intense. Signaled by bells and whistles, they learned to perform their tasks mechanically. They settled into the monotony of factory life through conditioning, but ultimately, young women's hearts lived elsewhere—by the sides of their parents, at home.[25]

As emotionally wanting as these women might have been once separated from kin, they might have been better off than their sisters in domestic service because they did not endure factory life alone. Employees formed alliances in workshops and dormitories that supported their mutual interests. Such attachments, though perhaps not as enduring as their rural friendships, were, at times, strengthened by immediacy. Networks inside the factories offered protection for some but also held policies of exclusion. Ties between women frequently depended on familiar affinities, such as the proximity of hometowns, age, religion, and social background. These women's relations were not always affirmative. Workers' affiliations were inclusive as well as exclusive, with boundaries more often fluid than fixed. Alliances between co-workers not only empowered individuals but served to regulate shop floor and dormitory norms.

Although resistance has been perceived as external to dominant systems, the bargaining networks of factory women were determined by the very institutional structures that repressed their autonomy. Apart from the workshop and dormitory, classrooms and places of worship were usually built on company grounds. The Dai Nippon Spinning and Weaving plant in Yŏngdŭngp'o, for example, implemented daily two-hour classes on arithmetic and Japanese by the 1940s. Buddhist shrines and classrooms, like workshops, dormitories, and dining halls, were spaces of socialization. For a few moments here and a minute there, workers shared their chores, meals, and burdens, relieving their fatigue by confiding in friends. Ironically, the factory system engendered the alliances that facilitated women's association and collective action.[26]

Even in times of seeming tranquility, survival for factory women meant breaking the rules. Insubordinate activities such as smuggling cotton or raiding storage facilities usually involved the collaboration of several workers. They knew that severe beatings were consequences, but for many, the risk was well worth the reward. As told by Kang Pokchŏm, one afternoon, she and two other women of the Tōyō (Tongyang) textile plant in Yŏngdŭngp'o could no longer stand their hunger and planned to smuggle food. That night, they successfully invaded the kitchen storage room. Overjoyed upon seeing the gigantic tubs of food before her, Kang confessed that she jumped into one of the larger-than-life bins, wallowing around in white powder. Whether out of hunger or a momentary sense of victory, Kang claimed that "plain wheat flour (*milgaru*) had never tasted so good."[27] Such adventures, if accomplished without detection, earned the intruders much popularity and symbolic capital back in the halls.

Working women in colonial Korea, particularly those employed in the cotton textile and silk-reeling industries, worked, slept, and lived together. Thus, it is not hard to imagine how they assembled, discussed plans for changes, and acted for their common interests. Operatives of the same station or area, for example, alerted each other of approaching supervisors or helped tend one another's sections. Factory women also garnered influence through calculated compliance. While some testified they were refused breaks because of a supervisor's disfavor, others believed they received satisfactory or even special treatment because they were skillful workers. With discernable pride, Han Kiyŏng asserted that she was liked by her supervisors because she was an exceptional piecer. Under different circumstances, Han's obedience might have inspired jealousy, but as the youngest in her section, she was protected by her co-workers. Han's testimony suggests that compliance and resistance were not stances that were simply adopted or rejected, but processes that were constantly negotiated. Although seemingly defenseless, factory women exercised everyday forms of resistance—beating the system through collaboration. Weapons of the weak, such as "foot dragging, false compliance, dissimulation, feigned ignorance, pilfering, slander, arson and sabotage," helped young women endure their labors.[28] When tactics of resistance were exhausted, however, mill girls ran away.

## Empowerment Through Escape

Factories in colonial Korea, especially those housing young women, undertook every conceivable measure to contain their workforces. The dormitory system, schemes of supervision, and company policies of restricting familial communications were reinforced by surveillance teams who patrolled the premises twenty-four hours a day. In spite of these efforts, every night, women eluded guards, crept through factory grounds, and jumped over barbed-wire fences to freedom. Meiji silk reelers described the prevalence of the phenomenon with a saying: "The day may come when the cock ceases to crow but never the day when factory girls stop running away."[29]

Although domestic servants (*kasa sayongin* or *sigmo*) escaped individually, industrial workers rarely attempted to leave the factory grounds alone. Groups planning escape were small; generally no more than three or four employees ran away at a time. With hopes that one would be more familiar with the terrain, or the other would have relatives closer by, they relied on shared support in planning and executing their escape. As with other aspects of factory life, women did not risk their livelihoods without each other.

Industrial settings made running away (*t'alch'ul*) difficult, forcing women to look toward distant relatives and local sympathizers for help and protection. As narrated by an account in the *Tonga ilbo* published in March 1931, girls between fourteen and fifteen years of age from Yŏngil and Yŏngch'ŏn frequently ran away from the Saengsa Company mill in Taegu. In a number of cases, absconders "went to houses in the outlying areas surrounding the factory and asked for refuge." Nonetheless, supporters of runaways were not always local families. The parents and relatives of factory girls often traveled to the mills with hopes of bringing their daughters home. According to an October 1926 newspaper report, upon hearing that women workers of the Yamajū (Sansip) silk-reeling factory jumped over the walls each night, the parents and relatives of employees gathered at the mill's entrance and waited for the girls to assist in their escape. Fearing for their children's safety, parents tried to climb over the gate in the hopes of finding their daughters. The company did its best to expel these relatives, "to the extent that guards would physically hit the parents and, in many cases, police intervention was necessary to keep these people away."[30]

In times of labor scarcity, these young women found work in other factories with the help of recruiters and managers who turned blind eyes. But as the networks between corporations and the police strengthened throughout the 1930s, runaway workers were caught and treated as criminals. As Yi Chaeyun, a former employee of the Kanegafuchi mill in Kwangju, recalled, when workers were discovered missing, "there was havoc in the factory." Supervisors "ransacked the dormitories," interrogating friends and co-workers: "Everything and everyone was scrutinized. All attention was placed on finding the missing employees." Searches did not stop at the company gates. Accompanied by hounds, patrols surveyed the outlying areas, venturing to more remote rural villages if needed. Frequently, the imperial police led external investigations. So, even if the girls made it back to their homes, authorities waiting there found them.

Running away resulted in punishment. When located, absconders were taken back to the factory compounds where, most often, supervisors took direct punitive measures. If the police intervened, runaway women could have been criminally sentenced and imprisoned for contractual violation. While imprisonment would seem a less favorable option to present-day observers, as Kang Pokchŏm narrated, days served in penitentiaries (*hyŏngmuso*) were "holidays" in comparison to the days of exhaustion in the factory. Kang conceded, "The only drawback was that the toilet was inside the cell. I could not get used to that. But I was better fed in jail, so I did not care." When returned to the mill, beatings, solitary confinement, and other reprimands awaited the runaways. Although factory women regularly endured punishment for misdemeanors, the penalties for running away were the most severe. In a number of cases, workers were exampled as wrongdoers—stripped of their clothing and made to stand on platforms wearing derogatory signs. At worst, failed getaways ended in sexual assault or rape. As Yi Chaeyun remembered, when a girl who tried to run away was caught, the outcome was brutal. She was taken to the supervisor, who would say: "You have two options: you can now either take all of your clothes off and run around the compound or, better yet, spend the night here."[31] Foundered escape plans resulted in violence. Yet, as the pervasiveness of sexual abuse during night shifts indicates, staying inside the gates did not prevent assault.

To understand why women ran away given the severe consequences, both emotive and material circumstances must be considered. Upon police interrogation, these women divulged that they could no longer bear the "severity of labor." Related to their lamentations over shop floor settings were problems with supervisors. Two separate accounts of runaways from the Michize filature in Kwangju noted that women ran away to escape beatings and verbal abuse. Similarly, when three women who escaped from the Katakura silk-reeling factory in Hamhŭng were asked why they ran away, they replied: "The restriction of freedom is worse than a prison and the severity of labor is unequal to any other." The absconders also asserted they could "no longer bear the ruthless misconducts of the overseers."[32]

After escape, women did not first seek employment but refuge. They frequently traveled for days and nights, finding sanctuary here and food there, but their final destination was home. The burden of their labors and their homesickness were not isolated as their emotional and corporeal conditions reinforced each other. Their displacement heightened their fatigue, and the difficulties of unfamiliar work intensified their yearnings for familiarity. The faster the threads broke and the harsher the beatings by supervisors, the more women wanted to "be at the side of their parents." These wishes were mutual, for as soon as a woman mentioned her desire to go home, "floods of tears" soon filled the dormitory rooms.[33]

Nevertheless, factory women's escape methods demonstrate the collective and creative abilities of workers in colonial Korea. Absconding was far more common than previously depicted, and as former mill operatives testify, more nights out of the year were filled with searches than not. In the larger plants, several runaways per night were customary. The fact that running away was just as much a part of factory life as corporate structures of discipline affirms that resistance, defiance, and escape were intrinsic to the industrial system. Everyday modes of survival, such as smuggling cotton and food, as well as radical retaliation, including pillaging storages and running away, required workers' cooperation. All of these activities demanded inventive planning and organization, or as Kang Pokchŏm described, "collaboration." In order to break out, women did not simply "jump fences," but timed the rounds of patrols and distracted guards while others escaped. To abscond, Han Kiyŏng and two other girls at the Dai Nippon (Tae Ilbon) textile factory in Yŏngdŭngp'o pretended to play hopscotch near the fences to divert attention. As characterized by a contemporary mill operative, the factory was a workplace, "but at the same time, it [was] a prison." The moment that women entered the mill, they "lost their freedom."[34] A critical investigation of the records on and testimonies of factory women indicates that though their bodies might have been contracted, the "weak" were autonomous in the creation of their dreams and schemes—their ideas. Conceptualizing escape as a practical and symbolic beginning rather than an ending allows for the recognition of factory girls as women who, often long after their days at the mills, worked for wages throughout the twentieth century. For better or for worse, fleeing the factory ended one form of labor and generally started another. Still, inasmuch as escape offered alternatives, running away might have been not a gesture of defeat but an assertion of self-determination.

*Summation*

In early twentieth-century Korea, nationalists, socialists, and feminists alike employed the image of the factory girl to underscore the adversities of Japanese colonial rule. Although scholars in late twentieth-century South Korea have provided more detailed information on colonial women workers, focusing on imperialist and capitalist resistance, many histories of colonial factory women end abruptly in the late 1930s, when repressive wartime policies curtailed labor activism. The guiding concern of generations of researchers has been what arguably occurred on one side of the peninsula and not the other. Thus, the writing of labor history has been "inevitably influenced by the desirability" of social revolution.[35] Due to the overriding interest in the revolutionary potential of women workers, little is known of what these women did before entering the factories or their accomplishments after they left the mills. Even less is known of factory women as individuals, and the emotional contents of their lives remain unearthed.

Because many scholars of colonial Korean labor defined consciousness with the larger intents of advancing nationalism and socialism, the identities of workers have been seen as invariable consequences of larger political and material realities. Nonetheless, a careful analysis of working women's interpretations of skill, labor, power, and alliance, as evidenced by not only their language but their behavior, invariably points to a frequently missed point: if instead of looking for universal markers of identification, researchers consider individual consciousness and labor's place in self-perception, then the place to begin is not with class but with work.

Stripped of its connotations, work was and remains, simply, the output of human energy that is both mental and physical. Although actions could have been mostly solitary or mostly social, an individual's effort was usually compelled by his or her roles and responsibilities in distinct social settings. In any given social formation, which activities "counted" as work were mediated by the cultural surroundings that prompted the work itself. The myriad of ways in which societies and cultures appropriate value to labor inform how individuals embedded meaning into work.

Work was not simply measured by its output. Rather, the customs and beliefs upheld by communities and households of interests, which predated capitalist mechanization, persisted after industrialization. Thus, while considering unpaid forms of labor is important for understanding the connections between labor and consciousness, it is paramount for an evaluation of women's work. The familial need for reproduction, child rearing, and household management was motivated by and, in turn, influenced the paid labors of family members. Therefore, the private and public spheres were not separate entities but fluid characterizations for the different roles and tasks needed to sustain the same economy. Although both men and women were affected by their familial, communal, and social obligations, women generally bore greater domestic burdens even after modern economic development. Exploring the lives of female factory workers sheds light on not only women's roles in capitalist economies but also the fluctuations of moral economies, or the social and cultural norms governing economic activity.

Labor disputes in Korea rose to conspicuous heights during the interwar years, but the outcome of the colonial labor movement, in at least one-half of the peninsula, would prove disappointing for those seeking reinforcement for coherent models of class formation. The actions of female factory workers in colonial Korea illustrates that like men, women engaged in labor demonstrations, but their contests of power were for neither the working class nor the nation. Men and women employed in dissimilar sectors fought for explicit reasons and separate interests. Like skilled male laborers in heavy industries, female operatives in cotton textile and rubber enterprises organized general strikes, halting production for months. Those employed in knitting factories gained leverage by engaging in citywide boycotts, whereas silk reelers of the same company or region often struck in rounds. Not just women but also men and children aided and resisted capitalist authority in contextually contingent manners, which varied greatly from metanarratives of ethnic or working-class solidarity. Although organized demonstrations displayed one form of agency, investigating workers' internalizations of skill, procedures of interpersonal empowerment, and modes of association confirms that strikes were simply extensions of daily practices of resistance. The working women of colonial Korea might not have actuated structural transformation, but through the common practices of calculated compliance, deception, resistance, and rebellion, mill girls kept capitalist authority in check.

Critics have maintained that women workers were unrevolutionary because characteristics of distinction, such as age and background, worked circumstantially to unite and divide. If consciousness was determined by labor activism, working-class consciousness in the colonial era was short-lived. Nevertheless, envisioning factory work as one of the multitude of tasks encountered throughout the life course indicates that working women of the early twentieth century did not identify exclusively with industrial labor as the rubric "working class" implies, but rather, held heterogeneous identities reinforced by their responsibilities and routines. The actions and articulations of factory women convey that, rather than working for the class or nation, women toiled for their families, communities, and close associations, suggesting that consciousness was shaped by the rhythms of daily life. Insofar as workers gained satisfaction by acquiring assets, whether material or symbolic, their motives were arguably more capitalist than socialist. Nonetheless, women's aims were not solely individualistic, because their aspirations were inseparable from those of their various communities of interests. In view of the fact that they failed to achieve socialist or nationalist revolution, the working women of colonial Korea might not have been revolutionary. But an intimate exploration of their contests of power affirms that they succeeded in fulfilling their personal ambitions, that is, in evoking revolutions from within.

# The Pacific War and the Life Courses of Working Women

On Sunday, December 7, 1941, at 7:53 A.M., 353 Japanese carrier-based planes launched a devastating surprise attack on the U.S. Pacific Fleet in Pearl Harbor, Hawaii. Several days after the bombing of Pearl Harbor and more than 6,000 kilometers northwest of O'ahu, near the Noryŏng mountains in North Ch'ungch'ŏng province in central Korea, Kang Pokchŏm encountered a young man on her way to school. Although Kang was only finishing the fourth year of her elementary education, she was expected to quit school and become a domestic servant because her family was poor. The stranger offered an alternative to servitude: she could simultaneously study and earn money. Thus, in December 1941, Kang Pokchŏm, like millions of Korean women before and after the war, transitioned from student to factory girl (*yŏgong*).

Escorted by her recruiter, Kang soon commenced work in the reeling division of the Yŏngdŭngp'o factory of the East Asia (Tōyō) Spinning and Weaving Company in present-day Seoul. Born in 1926, she had spent much of her childhood in the countryside. At the age of fifteen, she left her family and hometown for the first time in her life. Kang's journey from farm to factory was an experience shared by millions of Korean women who were mobilized for industrial labor during the Japanese colonial era.[1]

Whether Kang and her contemporaries knew it or not, their experiences were shaped chiefly by three watershed events preceding Pearl Harbor: Japan's annexation of Korea in 1910; its occupation of Manchuria, completed by 1932; and the Second Sino-Japanese War, which began with Japan's 1937 attack on southern China. The succession of conflicts termed the *Asia-Pacific War, the Pacific War,* or the *Fifteen Year War* started in the early 1930s with military aggression as well as with territorial and market expansion but affected few Koreans before the latter half of the decade.[2] The war against the Allied powers (1941–45) further fueled militarism throughout the Japanese empire. The August 1940 announcement of Japan's plans for protecting the Greater East Asian Co-Prosperity Sphere rationalized its mission for control over the so-called southern regions, including Burma, Malaysia, Indonesia, the Philippines, the Kuriles, Attu, Kiska, Kiribati, the Solomon Islands, Western New Guinea, and the Sunda Islands. Japan's quest for pan-Asian hegemony brought grief, suffering, and death to millions. The war also led to bureaucratic extension and accelerated economic development as well as transformations in the characteristics of education and employment in Japan and its colonies, namely Korea, Manchuria, and Taiwan. Perhaps more important, the war brought about enormous changes in the lives of ordinary people in these countries.

In Korea, modernization intersected with Japanese colonization. Within the first decade of annexation, the investment opportunities afforded by the First World War ushered in capitalist development on the peninsula. Expansion, however, was modest before Japan's 1931 invasion and occupation of Manchuria. Thus, though factories in early colonial Korea recruited young women directly from the countryside, by the late 1930s, schools, local patriotic organizations, and recruitment offices increasingly enlisted employees. After Japan's incursion into southern China in the summer of 1937, the Government-General of Korea actively redeployed both male and female workers to newly prioritized sectors such as iron and steel mining, manufacturing, and ship and aircraft building. The heightened production demands and labor deficits brought on by the Pacific War reconfigured the

Korean peninsula into an integral producer of Japan's "go-fast" imperial project, opening unchartered arenas of employment for Korean women, including jobs in heavy industries.[3]

The new order in East Asia, announced in November 1938, sought to unify the Far East under Japanese leadership. Ideological promotion corresponded with material growth throughout the 1930s as manufacturing priorities shifted from light to heavy industries. By the start of the Second Sino-Japanese War, the output of capital goods comprised almost half of the peninsula's gross domestic product. Economic development coincided with other features of modernization, such as population growth, urban migration, and the rise of an industrial workforce. By the late 1930s, the Korean population was growing steadily at annual rates of almost 2 percent. Simultaneously, the urban population burgeoned, from composing around 4 percent of the total population in 1920, to 8.4 percent in 1939, to 14 percent by 1945. These occurrences supported the proliferation of a mechanized workforce; whereas in 1931, about 1,200 medium- to large-scale factories employed 65,600 laborers, by 1939, approximately 212,500 laborers worked in nearly 7,000 large-scale factories. In the span of less than a decade the Korean industrial labor force had more than tripled in number.[4]

To date, research on factory labor in colonial Korea has predominantly focused on the activities of working men, who formed the majority of the industrial labor force. Less known is that from the outset of modernization, women composed over a third of the labor force in Korea. Throughout the 1920s and 1930s female employees made up the majority of the workforces in light industries such as food processing, rubber shoemaking, knitting, silk reeling, and cotton spinning and weaving. Although a handful of scholars have commendably investigated women who worked in the colonial era, focusing on nationalist and working-class activism, their studies of factory women end in the late 1930s, when officials clamped down on nationalist and labor activities on the peninsula.[5] Female factory workers of the colonial era thus appear as a group of women who rapidly rose and declined in significance; little is known of Korean women workers during the war years. By theoretically confining female employment to the textile sector, some historians of colonial Korea have implied that as state attention to textiles waned while interest in munitions and defense projects soared, the numbers of women in wartime industrial labor, as a whole, diminished.

Korean factory girls of the 1920s and 1930s are often portrayed as early victims of a capitalism that was brought on by Japanese imperialism. Interpretations of Korean women of the late 1930s and early 1940s continue this genre of historiography with highly politicized portraits of comfort women, or sex slaves for the Japanese military during the Second Sino-Japanese War. While this recent attention to comfort women has laudably exposed Japanese war crimes, it has also reinforced representations of colonial Korean women as powerless people.[6] The contemporary movement for the indemnification of former comfort women, whether deliberately or unknowingly, has linked wartime coloniality and femininity with sexual slavery. The fact that activists and academics use the Labor Volunteer Corps, otherwise known as *chŏngsindae*, and *comfort women* (*wianbu*; *ianfu*) interchangeably elides the diversity of women's wartime labor. Such misalliances erroneously suggest that *chŏngsindae* or *teishintai* was somehow synonymous with military sexual slavery, or comfort women.[7] *Chŏngsindae* translates literally as "offering-up-one's-body corps" but was both officially and commonly interpreted as "volunteer corps." Nevertheless, subjects of the Japanese emperor, including the people of Korea, Manchuria, and Taiwan, were mobilized as *paid* volunteers. Wartime recruits worked as factory operatives in Japan, Korea, Manchuria, and elsewhere in Japan's ever-expanding empire in the Pacific. Although *chŏngsindae* might have been used colloquially after the conflict to denote all women who served as wartime volunteers, more precisely, *chŏngsindae* refers to the institution formed under the Ordinance on Women's Labor Volunteers (Yŏja chŏngsin kŭllo ryŏng; Joshi teishin kinrōrei) issued in August 1944. The Women's Labor Volunteer Corps was prominent but was also only one of many patriotic labor leagues and campaigns formed during the war, particularly after Japan's attack on Pearl Harbor.

For historians of late colonial Korea, the dearth and disarray of materials concerning the Women's Labor Volunteer Corps and mobilization necessitates much piecework. Official statistical collection not only decreased drastically during the war, but much of what was recorded was destroyed. Apart from coverage by state-run newspapers, the *Maeil sinbo* and the *Keijō nippō*, and a handful of surveys and studies,[8] few prosopographic or narrative records concerning the composition and activities of the corps remain. Moreover, the scarcity of information on the specific factories women entered makes it difficult to gauge the geographic and professional scale of the corps. But the extant official sources, oral testimonies, and other materials, such as the records compiled by the Society for the Survivors and the Bereaved Families of the Pacific War, provide rare glimpses of the types of work women encountered.[9] Alongside the military accounts of the Second World War, the lesser-known stories of Korean factory girls reveal how ordinary women adapted to and in turn influenced the social, cultural, economic, and political transformations brought on by modernization and the two world wars.

This chapter on female labor in wartime Korea has several aims. First, I aspire to reveal the extent of Korean women's contributions to mechanized production in the Japanese wartime empire. Second, by outlining some of the central policies and programs that brought together companies and workers under the rubric of "imperial mobilization," I furnish examples of the scope of the war's social effects. An evaluation of the roles of schools, villages, and local organizations in late colonial Korea details how official policies and programs were institutionally executed. Third, I elaborate on some of the ways in which female labor recruitment was performed under the auspices of student campaigns in the immediate years before and especially after Japan's offensive against the Allies in the winter of 1941. Still facing a labor shortage in 1943, officials decided to recruit women for heavy industrial labor and aimed for the greater enlistment of women in the colonies. Describing how the Women's Labor Volunteer Corps mobilized Korean women for all types of war-related enterprises is the fourth objective of this chapter. Finally, I focus on the oral histories of female volunteers employed in the machine and machine tools sectors, specifically the operatives of the Fujikoshi steel factory in Toyama, Japan, to offer alternative renderings of working women in wartime Korea. In so doing, I hope to relay some of the heterogeneous outcomes of World War II in East Asia. Although this examination aims to expose some of the lesser-known effects of Japan's total war in Korea, it in no way attempts to undermine the suffering resulting from Japanese wartime atrocities.

Whether motivated by false promises, financial concerns, or the need for independence, young women such as Kang Pokchŏm entered the mills throughout the colonial era, and from the 1920s on, the numbers of factory girls rose at an unprecedented rate. These women—millions by the final years of the Second World War—redefined their perceptions, activities, and identities through their new work experiences. Though rooted in the conventions of rural turn-of-the-century Korean life, they deviated from tradition as they altered their inherited constructions of womanhood to fit their roles as industrial workers. While factory labor yielded little reward for some, others, sometimes spurred by their familiarities with wage work, eventually attained impressive degrees of economic and social autonomy. In the long trajectory of twentieth-century history, factory girls recruited during the Second Sino-Japanese War were among the first Korean women to diverge from their traditional roles, not only by entering the public sphere but by staying there, as they engaged in wage work throughout their lives.

## *The Imperial Mobilization Movement*

Although the first modern manufacturing projects in Korea commenced after WWI, prior to the late 1920s, most enterprises remained small. After the founding of Manchukuo, however, large-scale factories emerged as Japanese businesses sprawled further into the continent. The war against China proper fueled capitalist development and led to the transformation of managerial and productive methods that brought economy, society, and culture under official jurisdiction. In March 1937, the Government-General of Korea enacted the Law for the Control of Major Industries, giving the state greater control over the establishment and development of armament projects. By September 1937, through the Temporary Fund Control Law and the Temporary Regulations for Imports and Exports, officials assumed supervisory roles over finance and trade. In January 1938, the Mobilization of War-Related Industries Law permitted the Government-General to subsidize armament production. The breadth of state intervention in business, however, was uncertain until the Commission to Investigate Countermeasures for the Current Situation met in September 1938 to form new guidelines for economic development. The commission called for the expansion of sectors deemed essential to the war effort, increased recruitment, and the reorganization and training of the Korean labor force, as well as the conversion of small- to medium-scale factories into "subcontracting workshops" for large-scale munitions plants.[10] The Industrial Association Law of

September 1938, resulting from the investigation, allowed the Government-General to create bureaus that steered the distribution, manufacturing procedures, and production levels of all industries.

The broad contours of labor mobilization in wartime Korea have been characterized by three stages: group recruitment, 1939–41; official arrangement, 1942–43; and coercive enlistment, 1944–45. The National Draft Law, instituted in April 1939, paved the way for the voluntary recruitment system (*bōshu*), active from September 1939 to March 1942. Mobilization during this period was termed "voluntary" since only companies and private contractors were required to follow the Government-General guidelines for obtaining labor. Nonetheless, as W. Donald Smith contends, when appraised alongside contemporary policies, such as the "harmony-promoting" assimilation programs and the July 1940 directive to reinforce the police forces of Pusan and Shimonseki, the voluntary claim of this phase of mobilization loses its credence. In response to receding labor supplies, dwindling especially after Pearl Harbor, a government-managed (*kan assen*) system was launched in March 1942. To centralize labor conscription, the Government-General established the Korean Labor Association, which supervised enrollment and employment conditions. Through group recruitment, initiated in 1942, authorities set workers' terms. A formal contract period of two years was instituted, but in 1943, it was lengthened to two and a half years. Despite these and other vigorous efforts, the demand for workers escalated. Thus, in September 1944, officials instated a system of forced or coercive (*chōyō*) recruitment.[11]

State-led labor mobilization in wartime Korea began in April 1937 with the introduction of the Employment Promotion Policy. The program intended to recruit employees for heavy industries, including mining and construction, and to reform the productive procedures of these enterprises.[12] In September 1937, the National Spiritual Mobilization program was inaugurated, ideologically grounding civilian support. The recruitment campaigns of 1937 and the imperialization (*hwangminhwa*; *kōminka*) movement expressed the message of colonial authorities—Koreans were to serve as the promoters and defenders of the Japanese emperor with total devotion of mind, body, and spirit (Table 5.1).[13]

After the March 1938 declaration of the National General Mobilization Law,[14] labor recruitment assumed a greater role in social organization. The

TABLE 5.1
## Chronology of Labor-Related Orders for the War Effort (1937–45)

| | |
|---|---|
| 3.1937 | Law for the Control of Major Industries |
| 4.1937 | Employment Promotion Policy (Kan assen shokugyō seisaku; Kwanalsŏn sanggongŏp chejak) (Government recommendations for Korean laborers to specific industrial projects) |
| 9.1937 | National Spiritual Mobilization Movement (Kokumin seishin sōdōin undō; Kungmin chŏngsin ch'ongdongwŏn undong) initiated |
| 9.1937 | Temporary Fund Control Law and Temporary Regulations for Imports and Exports |
| 1.1938 | Mobilization of War-Related Industries Law (Gunju kōgyō dōinhō; Kunsu kongŏp tongwŏnbŏp) |
| 2.1938 | Laws Concerning Special Army Volunteers promulgated. Initiation of the Volunteer System (Shiganhei seido; Chiwŏnbyŏng chedo) for the Japanese Imperial Army |
| 3.1938 | National Mobilization Law (Kokka sōdōinhō; Kukka ch'ongdongwonbŏp) |
| 6.1938 | Task Concerning the Establishment of Labor Duties for Students (Gakusei seitō no kinrōhoshi sagyōshise ni kansuru ken; Haksaeng saengdo ŭi kŭllobongsa chagŏpsilsi ŭi kwanhan kŏn) |
| 7.1938 | Establishment of the Korean League for National Spiritual Mobilization (Kokumin seishin sōdoin Chōsen renmai; Kungmin chŏngsin ch'ongdongwŏn Chosŏn yŏnmaeng) |
| 7.1938 | Establishment of Monthly Patriotic Day |
| 8.1938 | Commencement of Monthly Air-Raid Drills |
| 9.1938 | Commission to Investigate Countermeasures for the Current Situation |
| 9.1938 | Industrial Association Law (Chōsen kōgyō kyōkairei; Chosŏn kongŏp hyŏphoeryŏng) |
| 9.1938 | Regulation Limiting the Hiring of New Graduates (Gakkō sotsugyōsha shiyō seigenrei; Hakkyo cholŏpcha sayŏng chehallyŏng) |
| 9.1938 | Regulation Limiting the Hiring of New Employees |
| 3.1939 | Regulation on Work Hours and Wages |
| 4.1939 | Task Concerning the Duties of Labor Leagues (Shūdan kinrō sagyōshise ni kansuru ken; Chiptan kŭllo chagŏpsilsi ŭi kwanhan kŏn) |
| 7.1939 | National Draft Law (Kinrō choyorei) instituted |
| 7.1939 | Special Volunteer Army Program initiated |

(continued)

TABLE 5.1 (*continued*)

| | |
|---|---|
| 7.1939 | Document Concerning the Migration of Korean Workers to Japan (Chōsenjin rōmusha naichi ijū ni kansuru ken; Chosŏnin nomuja naeji iju e kwanhan kŏn) |
| 8.1939 | Regulations Limiting the Hiring of New Employees |
| 9.1939 | Voluntary or Group Recruitment System (Boshu) instituted |
| 10.1939 | Ordinance on the National Draft (Kokumin chōyōrei; Kungmin chingnyŏngnyŏng) initiated procedures for recruiting civilian workers for the military |
| 1.1940 | Regulation on Employment Offices (Chōsen rengyo shōkaisho rei; Chosŏn chigŏp sogaeso ryŏng) |
| 8.1940 | Termination of All Newspapers (except the Government-General-sponsored Maeil sinbo and Keijō nippō) |
| 10.1940 | National General Mobilization Movement (Kokumin sōryoku undō; Kungmin ch'ongnyŏk undong) |
| 10.1940 | National All-Labor Movement (Kokumin kairō undō; Kungmin kaero undong) |
| 4.1941 | Rationing of Daily Commodities and Provisions begins |
| 6.1941 | Association for Korean Labor Affairs established (Chōsen rōmu kyōkai; Chosŏn nomu hyŏphoe) |
| 12.1941 | Patriotic Labor Association Law (Kokumin kinrō hokoku kyōryoku rei; Kungmin kŭllo poguk hyŏmnyŏngnyŏng) and the formation of Patriotic Labor Corps for National Support (Kinrō hōkokutai; Kŭllo poguktae) |
| 12.1941 | National Registration of Youth and Adults (Seisōnen kokumin tōroku; Chŏngjangnyŏn kungmin tŭngnok) |
| 1.1942 | Labor Regulation Law (Rōmu chōseirei; Nomu chojŏngngnyŏng) |
| 3.1942 | Government-Managed Recruitment (Kan assen) System begins |
| 7.1943 | Provision for the Establishment of Wartime Student Organizations |
| 8.1943 | Initiation of the Volunteer System for the Japanese Imperial Navy |
| 10.1943 | Countermeasures for Labor Enforcement (Rōmu kyōka taisaku yōzuma; Nomu kanghwa taech'aek yogang) |
| 3.1944 | Provision for Emergency Student Mobilization (Gakudōdōin hijōsochi yōzuma; Haktotongwŏn pisangjoch'i yogang) |
| 5.1944 | Provision for Emergency War Operations (Kessen hijōsochi yōzuma; Kyŏlchŏn pisangjoch'i yogang) |

(*continued*)

TABLE 5.1 *(continued)*

| | |
|---|---|
| 8.1944 | Student Labor Law (Gakudō kinrōrei; Hakto kŭlloryŏng) |
| 8.1944 | Women's Labor Volunteer Law (Joshi seishin kinrōrei; Yŏja chŏngsin kŭlloryŏng) |
| 9.1944 | System of Forced or Coercive Recruitment (Kyosei choyo) installed |
| 4.1945 | National Labor Mobilization Law (Kokumin kinrō sōdōinrei; Kungmin kŭllo ch'ongdongwŏllyŏng) |

plan sought to amass and optimize the "natural and human resources" of Japan, Korea, and Taiwan (Article 1). The Japanese government reserved the right to investigate the labor capacities of all imperial subjects (Article 21) and to conscript individuals for war-related labor (Article 4). The policy allowed for the "imperial assignment" of all potential workers (Article 22). The state also retained the right to control the labor conditions—the assignment, type of work, period of stay, wage, and dismissal—of each worker (Article 6).[15] Whereas the National Mobilization Law set the legal precedent for drafting troops and workers, the Korean League for National Spiritual Mobilization, formed in July 1938, provided an official bureau to micromanage recruitment. National and regional leagues composed the higher branches, but provinces, cities, counties, townships, villages, and streets comprised the lower administrative units. Factories, schools, businesses, banks, stores, and units of ten families in residential areas formed patriotic associations, which reinforced other enlistment campaigns. By June 1939, there were 347,728 patriotic associations throughout Korea.[16] The 1939 promulgation of the Ordinance on the National Labor Draft gave labor conscription imperial office, theoretically allowing for the total organization of the Korean populace.

To fortify the programs of the late 1930s, the 1940 Regulation on Employment Offices authorized the building of nine new centers for public employment, including branches in Kyŏngsŏng, Pusan, Pyongyang, Taegu, Sinŭiju, and Hamhŭng.[17] In March, the Labor Affairs Bureau of the Government-General began an extensive investigation of labor resources on the Korean peninsula, resulting in the Association for Korean Labor Affairs, which further consolidated the procedures of workers' enrollment and management.

In October 1940, the National Mobilization Movement was launched. The resulting body, the Korean League for National Mobilization, acted as a Korean version of the Imperial Rule Assistance Association.[18] Also in 1940, the All-Labor Movement commenced, buttressing recruitment programs already in place. From the offices at the top to neighborhood associations at the bottom, Korean society became integrated with the official administration.

As Japan's offensive widened with its attack on Pearl Harbor, bureaucratic promotion and recruitment for the war effort intensified. The Patriotic Labor Association Law of December 1941 introduced another vehicle for mobilization through the formation of Patriotic Labor Associations. A 1941 study of seventeen companies in the building and construction industry reveals that over half of their employees were recruited through Patriotic Labor Associations. Also in December 1941, the National Registration System, whereby Korean men and women between twelve and forty years of age had to register with the government for potential enlistment into war-related services, gained institution. All men aged sixteen to forty and unmarried women between sixteen and twenty-five had to enroll for imperial service.[19] The National Registration System furnished officials with information on pools of available workers who were identified, recruited, and assigned to specific tasks or projects.[20]

Amid a time of great uncertainty, the state sought to inculcate constancy through ritual. In July 1938, a year after Japan's invasion of Shanghai, officials announced the inauguration of the Patriotic Corps of Industrial Workers (*sangyō hōkokutai renmei*), or the *sanpō* movement, as a corporate subdivision of the National Spiritual Mobilization Movement. Designed to streamline production, it established a central secretariat in Tokyo and thousands of branches in factories, mines, and construction sites throughout the Japanese empire. Effectively a state-run federation of unions, the corps encouraged regular meetings between laborers and managers, stressing the value of each employee's contribution regardless of rank. After Pearl Harbor, however, its philosophy shifted from "enterprise as one family" (*jigyō ikka*) to the propagation of the "imperial work ethic" (*kōkoku kinrōkan*), which emphasized punctuality, efficiency, productivity, and sacrifice instead of mutual respect.[21]

The *sanpō* campaign installed patriotic units (*aikokuhan*) in industrial sites throughout Japan, Korea, and Manchuria, which met regularly and

*Plate 5.1*   Rural women mobilized for agricultural labor during the Pacific War (1940s).

*Plate 5.2*   Patriotic women's associations in late-colonial Korea (1940s).

*Plate 5.3*   Fencing practice in a school for girls (1940s).

engaged in public observance. From July 1938 onward, the first of each month was celebrated as "patriot day." The official ceremony began at "10 A.M. with workers bowing towards the Imperial Palace in Tokyo." Farmers, factory workers, students, and public officials alike prayed for the fallen, sang the Japanese national anthem, and heard lectures celebrating austerity and suffering as hallmarks of imperial subjectivity. Afterward, attendants sang the "patriot's marching song" and recited the "oath as subjects of the imperial nation" (*kōkoku shinmin no seishi; hwangguk sinmin ŭi sŏsa*):

1.  We, the subjects of the imperial nation, pledge to serve his majesty and the country with loyalty and sincerity.
2.  We, the subjects of the imperial nation, shall trust, love, and help one another to bolster our unity.
3.  We, the subjects of the imperial nation, shall endure hardship and cultivate our strength so that we can exalt the imperial way.[22]

At the end of the ceremony, participants pledged their allegiance to the emperor, shouting "Banzai!"[23]

The imperial mobilization movement, by bureaucratizing industrial production, quickened the momentum of economic development and reshaped the procedures of labor recruitment, placement, and management. Due to inconsistencies and gaps in remaining wartime sources, however, estimates for the numbers of mobilized Koreans vary. According to a 1947 Ministry of Finance report, accounting for the total number of Korean workers between 1934 and 1945, over seven and a half million Koreans were recruited for labor in Korea, Japan, Manchuria, and occupied territories in China. Approximately 4.8 million workers were mobilized for war-related projects in the Korean peninsula alone; the remaining migrated to China, Japan, Manchuria, and Southeast Asia.[24] With heightened wartime demand, light industries grew in size and scope and heavy industries multiplied. War preparations affected not only mechanized production but also agriculture, household manufacturing, trade, communications, education, and culture. According to the statistical yearbooks of the Government-General of Korea, the ratio of heavy industrial output to total production surged dramatically throughout the 1930s, from 23 percent in 1930, to 37 percent in 1935, to 50 percent in 1940.[25] From 1937 to 1943, the number of workers in construction enterprises grew from 161,499 to 380,000 and, in mining, from 166,568 to 280,000. Manufacturing likewise expanded; between 1937 and 1943, 155,000 new workers entered medium- to large-scale factories. Alongside material change, from 1937 to 1945, the procurement of human resources in Korea became institutional. The people embedded in public institutions, such as armies, schools, employment agencies, and neighborhood societies, became the most pervasive weapons of war.[26]

## Education and Labor Mobilization

The public education system in Korea was introduced by the Government-General in 1911. Attendance, though not widespread initially, grew by 1933 to include 20 percent of the elementary school–aged population, who were enrolled in more than two thousand ordinary schools. By 1940, the public school system had registered 50 percent of all Korean school-aged children.[27] Kim Chŏngmin, who attended public school in the last decade of colonial rule, explained that even less-well-to-do families sent their children to school because failure to enroll children invited "trouble from the Japanese authorities."[28]

Colonial education relied on the rhetoric of Confucian paternalism, but more instrumentally, schools were the training and selecting grounds for future civilians, soldiers, producers, and reproducers. One of the earliest recorded projects for student mobilization in Korea was the Task Concerning the Establishment of Labor Duties for Students, launched in June 1938. According to a brief report in the popular *Tonga ilbo*, in one task fourth-grade students from Kyŏnggi province assembled for ten days during intersession to assist in munitions manufacturing. Like classrooms, student apprenticeships were gender segregated and gender specific. As a part of the task, "boys worked in construction projects whereas girls cleaned Shintō shrines and sewed supplies for the military."[29] The Patriotic Labor Association Law of 1941 effectively turned students into workers as it reorganized schools into productive units.

After Pearl Harbor, projects for student recruitment became massive in scale. In July 1943, the Provision for the Establishment of Wartime Student

*Plate 5.4*   Wartime students write "Do away with rats" (1940s).

Organizations restructured all secondary and vocational schools for labor enrollment. The Provision for Emergency Student Mobilization, initiated in March 1944, enlisted students for war-related projects according to their academic concentrations. Female students in nursing schools, for example, worked in military hospitals.[30] Despite some degree of specialization, students were most frequently assigned to munitions plants by the last years of the war. In May 1944, as a part of the Provision for Emergency War Operations, 360 male and female students assembled for work in a military arsenal in Inch'ŏn. Although student projects recruited youth for short-term labor, the Student Labor Law, instated in 1944, organized groups of students into corps (*tae*; *tai*) to work in munitions projects for over a year or longer.[31]

Laws for the centralization of production and labor enlistment were sustained by ideological campaigns, deeply penetrating every layer of society, including the local, the familial, and the personal. Promotions such as the Patriotic Labor Movement of 1941 assembled the Korean populace into neighborhood units that buttressed more explicit mobilization projects. The ritualistic counterparts of corporeal labor stimulated society in diverse but equally provocative manners. Ceremonies including Monthly Patriot Day, and the institution of martial drills such as fencing and practicing the use of bamboo spears in schools, exhibit the degree of state infiltration into public, private, and cultural life. The war machine incorporated Koreans of all ages: young elementary school students practiced propagandistic calligraphy and sewed so-called comfort bags for imperial soldiers in the front lines, whereas older Korean youths were drafted for military service and enlisted for factory work.[32]

## The Women's Labor Volunteer Corps

Sources on Japanese women account for the activities of those recruited from August 1944, or January 1944 at the earliest, when proposals for the Women's Labor Volunteer Corps (Yŏja kŭllo chŏngsindae; Joshi kinrōteishintai) were first discussed in the Diet. Written evidence does not confirm whether women were mobilized under the auspices of the Women's Labor Volunteer Corps before 1944. Nevertheless, the government actively pursued female recruits before the enactment of the ordinance through other campaigns such as those targeting students. Although authorities enlisted women under an assortment of programs in the late 1930s and early 1940s, the critical labor shortage felt by 1942 forced the creation of an office for the sole purpose of mobilizing women. References to female student workers appear as early as 1941, but an exclusively female labor organization did not exist until 1943, when the Department of Public Welfare announced a policy to enlist middle school graduates for the Labor Volunteer Corps.[33] Officially, the Women's Labor Volunteer Corps called for students and school-aged girls to serve in war-related enterprises for one year, although, depending on individual and familial circumstances, some women worked for much longer.

Factory girls in colonial Korea were usually from impoverished backgrounds, but literature on the Women's Labor Volunteer Corps suggests that wartime volunteers were unlike their prewar predecessors. In describing a new plan for female recruitment, authorities asserted that the project would draft members of the leisured middle classes. In 1943, a Korean commentator, Yu Kwangnyŏl, argued that young women called to become volunteers should consider the offer "honorable," because the group represented the best qualities of "model" employees and patriots.[34] Particularly vaunted was the corps' acceptance of students who were thought to be socially and culturally connected to patriotic, labor, and women's groups already active in local society. In official promotional literature, however, acceptable candidates were named as "recent graduates of elementary and middle schools *as well as* unmarried women over fourteen years of age."[35] Not only were standards for admission low, the wartime government also restricted the hiring of women between the ages of twelve and twenty-five as domestic servants, thereby outlawing private competition for their recruitment. The combination of these low standards of admission and restrictions on private hiring practices testifies to the social and economic breadth of those drafted for work. Cross-referencing the corps' guidelines with firsthand testimonies exposes that the middle-class composition of the corps prevailed only in official rhetoric.

Although many Korean workers attained skilled and technical positions during the war years, the social and economic distribution of the Korean populace remained relatively unchanged throughout most of the colonial era. Tenants, agricultural wage workers, and industrial laborers composed more than three-quarters of the population by the end of the 1930s, just as similar proportions were peasants in the first two decades of imperial rule. Yet, the fact that by the end of the colonial era half of the school-aged children were enrolled in public education verifies that families earning less than the middle-class average sent their children to schools. Corresponding to women's participation levels in manufacturing and agricultural wage work, female students formed about 30 percent of the total in primary schools.[36] Furthermore, once women started schooling, they were more likely to continue, as females accounted for almost half of those enrolled in secondary schools.[37] The last years of the Pacific War were times of opportunity not only for middle-class Koreans but also for determined young women in the lower classes. While the corps enlisted students, its members were not necessarily the daughters of middle-class parents. In Japan and

Korea, young women already working in mills or as domestic servants volunteered for wartime labor because the corps provided higher wages.[38] Images of the middle-class volunteer, though targeted at middle-class Korean supporters of Japanese governance, also affected the ambitions and behaviors of lower-class youth.

The variety of employment offered through the corps was more diverse than ever. Women worked as operatives in "mining, electricity, communications, machinery, aircraft manufacturing, shipbuilding, chemicals, ceramics, carpentry and building and construction."[39] Indeed, the war required ever-increasing quantities of capital goods and numbers of laborers for their production. By 1945, those employed in construction, mining, and manufacturing comprised 20 percent, 15 percent, and 28 percent, respectively, of the wage-earning labor force in Korea. Women were most prominent in manufacturing but were also employed in smaller numbers in mines and construction sites. Although the war effort placed emphasis on heavy industries, textile production remained integral to the economy, and women filled these positions in more extensive, large-scale spinning and weaving companies. In addition to light industries, female employees were prominent in the machine and machine tools sectors as well as in airplane manufacturing. Few women employed in heavy industries were placed in the highest technical positions. Still, the fact that female volunteers engaged in heavy industrial labor indicates that Korean women encountered a greater range of employment opportunities for the first time during the Pacific War.

Candidates for the Women's Labor Volunteer Corps were selected in many ways. Teachers and school administrators recruited students, district leaders approached youth in cities, and village headmen frequently helped mobilize young women in the countryside. As in the schools, volunteers were frequently chosen through lottery systems in which individuals were randomly selected from all of the eligible enlistees in the village. Alongside local drives, company contractors and state officials assembled youth.[40] In one reported case, a fourteen-year-old domestic servant joined the corps after a friend told her of the wages offered by the corps.[41] Women also responded to advertisements on their own, as regional employment offices and training centers became widespread by the early 1940s.

The kind of work a young woman performed as a volunteer for the corps relied on her background and level of education. Advertisements called for students with proficiency in Japanese and other skills, but facing a dire labor shortage, the corps refused few applicants. Furthermore, wartime legislation placed virtually every business under state authority. Therefore, young women with schooling were granted semi-skilled positions, whereas those without educational or employment experience were placed in jobs traditionally considered unskilled, such as apprentice work in textile factories.[42]

According to the registry of the Society for the Survivors and the Bereaved Families of the Pacific War, the corps stationed women across East Asia. The largest armament projects were based in Japan, but throughout the 1930s machine and machine tool–making plants were built in Korea. Thus, female volunteers were employed in factories producing military equipment in cities such as Pyongyang and Wŏnsan. Women assigned to the textile sector usually worked in Korean factories, but volunteers also went to spinning and weaving plants in Japan.[43] Nevertheless, according to the society's records and contemporary reports, war-related heavy industrial projects, such as machine and machine tools manufacturing as well as aircraft and shipbuilding, also sponsored the women of the corps. Among these enterprises were the Fujikoshi steel factory in Toyama, the Mitsubishi airplane factory in Nagoya, the Yamada machinery factory in Kyūshū, a shipbuilding plant in Nagasaki, and another in Toyama. Female volunteers from Korea were sometimes situated even further. Across China and Southeast Asia, for example, women registered with the society were taken to industrial sites in Shanghai and Singapore.[44] (See Figure 5.1 for a map of East Asia.)

## Women Workers in the Machine and Machine Tools Industry

In spite of the scarcity of comprehensive information on Korean women who worked in the late 1930s and early 1940s, detailed oral and published sources concerning the Fujikoshi steel factory in Toyama, Japan, provide a rare opportunity to examine the experiences of women workers in a single setting.[45] Established in 1928, the Fujikoshi Steel Company in Toyama specialized in the production of steel goods, machine tools, and airplane parts. From its founding, Fujikoshi employed both male and female workers as Japanese women entered the heavy industries earlier than Korean women.

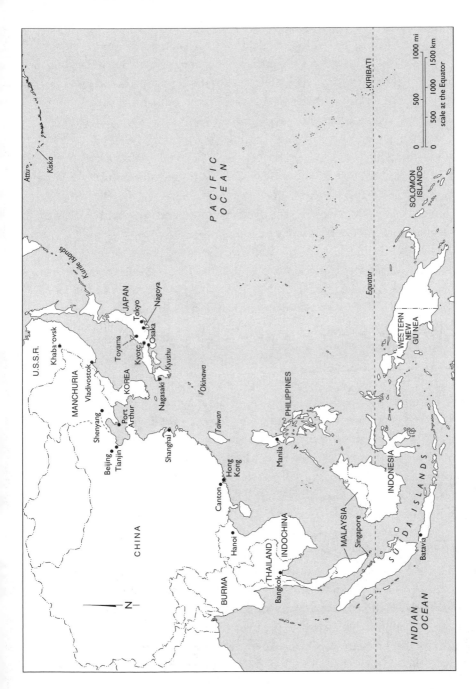

*Figure 5.1* Map of East Asia (1945)

Employing more than 36,000 workers by 1945, the plant was one of the most advanced in its technology, scale, and labor standards. Among other facilities, Fujikoshi boasted of a wind orchestra, an in-factory cinema, and an athletic field, built in 1939, which spread over 7.2 square kilometers. The factory compound had within its gates three large canteens that seated six thousand workers, plus medium-sized canteens and small stores. Each dormitory was equipped with bathing facilities and lavatories that suited five hundred residents. In 1940 a fully equipped hospital replaced the infirmary.[46] Because of its scale and progressive conditions, the factory was acclaimed in newspaper reports, and oral testimonies verify the details. According to Kim Chŏngmin, the factory compound was "unbelievably huge. You could not see from one end to the other."[47]

The Fujikoshi Steel Company had two kinds of plants: steel-making mills and tool factories. Although a plan to build Fujikoshi plants in northern Korea was launched in March 1945, volunteers were taken to the factory in Toyama. Korean women were involved in all forms of tool making with numerous types of machinery, such as planing, shaping, and milling machines. But the majority worked in the lathe (sŏnban; senban) operation, cutting machine parts and tools such as ball bearings. Before entering the shop floor, women underwent six months of group training. Upon completing the course, they performed shop floor tasks, including lathe operation. Because the young women were too small for the machinery, they worked standing on rectangular stools to reach the controls. All employees at the plant wore uniforms, and women with long hair wore caps for safety. Although owners and officials privileged the value of lathe operation over that of textile work, turning controls while standing still seemed less strenuous than running back and forth piecing yarn at lightning speed. In contrast to the tales of exhaustion in textile factories, Kim Chŏngnam said, "the work itself, though uninteresting, was easy."[48]

Regional ties between workers remained salient markers of association and distinction as Koreans from the same regions were housed together. Nonetheless, within a large and highly stratified system of working and living, the alliances of Fujikoshi operatives diverged from women's ties in smaller, domestic factories where employees were recruited from just a few areas. Although operatives recalled friends and Korean elders, they also spoke of their dormitory parents, called "residence mothers and fathers." The dormitory parents and the dormitory supervisor, usually in their late teens and early twenties, were seen as the guardians of the boarders. As Kim Chŏngnam recalled, the dormitory mother felt "real affection for the girls. She sent us off in the morning saying, 'Have a nice day,' or some other friendly expression." The dormitory father walked the girls to the plants in the morning and back to the halls in the evening. Wearing matching uniforms the women marched "like soldiers" and sang the company song on the ten-minute-long journeys.

Once they reached the plants, they headed to their designated stations, whereupon the supervisor came and "lifted the girls onto the platforms." He would give an encouraging pat on the head and say something like "work well today." Once there they shaped small pieces of metal into machine parts from 7 A.M. to noon. At noon they stopped for lunch. Like residence halls, factory canteens were systematized hierarchically. All of the women were assigned to one another as sisters, and, according to Kim Chŏngnam, the housed Korean workers had older Japanese sisters who were day workers. The Japanese sisters packed their younger sisters' lunches from home. The older sister prepared meals for the younger, and thus, as Kim explains, whether "you were hungry or full depended on the sister you met." Fortunately, Kim's elder sister was the daughter of a well-known Buddhist monk. She exclaimed, "So, imagine the kinds of treats I received." Others, however, were not so lucky. Kim recalled that her friend met "the daughter of a poor tenant." In spite of cultural dissidence, Kim maintained that they were "like real sisters. They looked out for us and provided for us, as though there was no difference in ethnicity or kin."

Women employed in machine and machine tool–making enterprises, particularly the operatives of the Fujikoshi factory in Toyama, experienced ritualized work routines and lifestyles. The day began at 5 A.M. when women woke to the morning greeting amplified through the factory broadcast system. Between 5 and 5:40 workers washed, dressed, made their beds, and cleaned their rooms. Afterwards they recited morning greetings and followed the calisthenics instructed by radio. From 6 to 6:50 residents engaged in morning worship, recited the empire narrative (*hwangminhwa sŏsa*; *kōminka joji*), greeted their dormitory parents, and ate breakfast. They walked from the dormitory to the factory and started work at 7 A.M. The women rested for lunch at noon and had an afternoon break at 3 P.M. Their shift ended at 5 P.M., and after clearing their stations, they returned to the dormitories by 5:30. From 5:30 to 7:40, they washed, changed clothes, ate dinner, and rested. From 7:40 to 9 they swept their rooms and cleaned, recited roll call, and greeted their dormitory parents for the evening. The lights were turned off at 9 P.M.[49]

Fujikoshi operatives worked from 7 A.M. to 5 P.M., but lunch and afternoon breaks took more than an hour combined. Informants recalled only eight hours of work instead of nine, indicating that the hours did not pass by too slowly. Varying from the schedules of the cotton mills of the 1920s, machinery workers were given rest days once a week instead of twice a month. Distinctions of industry and locale might have caused these deviations in labor standards, but cross-referencing the accounts of Fujikoshi women with those of workers in Korean mills indicates otherwise. Those employed in older Japanese factories in Korea founded in the late 1920s and early 1930s, such as the Kanegafuchi textile factory in Kwangju and the Tōyō textile factory in Yŏngdŭngp'o, endured twelve hours of work and had holidays only twice a month. Nevertheless, those employed in large-scale factories established in the late 1930s and early 1940s, including the Dai Nippon Spinning and Weaving factory in Yŏngdŭngp'o, encountered conditions similar to those at Fujikoshi. Although Han Kiyŏng was employed in a textile factory in Seoul, she only worked "eight or nine hours a day and had one rest day per week."[50]

For Allied and Axis nations alike, World War II provided unprecedented opportunities for testing technological and productive innovations, as well as for exploring new scientific methods of social experimentation and discipline. Whereas news of the First World War spread through print in days, by the late 1930s, international events, blazoned through radio, reached ordinary people in minutes. The broadcasting network established in Korea in the 1920s allowed officials to communicate directly with civil servants, students, teachers, workers, and farmers in the countryside. At the Fujikoshi factory, the radio introduced new modes of social conditioning. For workers and other civilians, routines, lifestyles, and even mentalities became increasingly synchronized and militarized as the intensification of conflict by the 1940s brought on rations, air drills, and days spent in bomb shelters.

During a time of turmoil, propaganda gained breadth, intensity, and totality, infusing popular culture with wartime ideology. The language of the spiritual mobilization and imperialization movements aimed for the total reconditioning of Japanese and colonial populations alike. Whereas the National Spiritual Mobilization Movement endorsed the giving of one's spirit to the emperor's cause, the Labor Volunteer Corps promoted physical labor as an expression of patriotism. Wartime propaganda combined the aspirations of the private and public realms of life by encouraging married women to serve their husbands by replacing their labor in the farms and fields and recruiting unmarried women for factory work to serve the emperor. In the official press, women of the corps were modeled as superlative patriots and idealized as the "flowers of Japanese-Korean unity" (*naesŏn ilch'e ŭi kkot*).[51]

Some perceived their duties as public, but most Korean youth enlisted to work for more personal reasons. Although education prepared the privileged for further training, schools led the poor into the labor force. Not just patriotism but financial need fueled the volunteer programs in colonial Korea. During a time of dramatic transformations, numerous youths saw volunteering as a means for both material and personal improvement. For authorities, industrial labor provided an outlet for women to shoulder their economic and moral responsibilities for the empire rather than to empower themselves as individuals. Yet students and other unmarried rural women became more self-reliant through the unparalleled experience of wage work.

*Summation*

Among the myriad of forces that affect modern society, warfare is "perhaps the most prevalent, possibly the most calamitous and probably the most consequential." Japan's invasion of Manchuria in 1931 and southern China in 1937 prompted technological advancement and accelerated mechanization on the Korean peninsula. War in Asia and the Pacific galvanized a centralization of authority that blurred the boundaries between individual and collective rights and liberties. War in late colonial Korea invoked social discipline by rigorously guiding daily life. The values and motivations of civilians became closer to those of militarists, and within this environment, a new social hierarchy emerged. As Gordon Wright describes, total war twisted and blocked the normal channels of social mobility "as by an earthquake" and wrenched open new channels for previously obscure people.[52] Modern warfare weakened social boundaries, but in early twentieth-century Korea, the immediacies of the Second Sino-Japanese War nearly flattened society.

The Pacific War confirmed that crisis (*wigi*) and opportunity (*kihoe*) were indeed coupled by timing (*ki* or *gi*). But how did historical timing influence those arguably quadruply disenfranchised in prewar society—colonized, poor, young women? War mobilization in late colonial Korea sought to turn young men and women into soldiers and workers. By infiltrating schools and communities, Japanese officials and Korean middlemen aggressively implemented policies and programs for extended human resource acquisition and material production. Two significant consequences of war mobilization for women were the surge of patriotic associations such as the Women's Labor Volunteer Corps and the rise of female employment in heavy industries including machine and machine tools manufacturing. These outcomes attest to young rural Korean women's contributions to wartime production. But how did factory work mediate the beliefs, actions, and identities of individuals after their service?

This exploration into female factory labor in late colonial Korea confirms that administrative developments, wartime policies on material output, and advances in human resource management as well as education fueled labor recruitment during the war. The simultaneous explosion of patriotic associations such as the Women's Labor Volunteer Corps and the proliferation of women workers in the machine and machine tools industry further illustrates how wartime policies broadened the ranges of female employment. Testimonies detail the expansion of women's work throughout the colonial era and the extent to which the terms and conditions of factory labor evolved during the war. More interestingly, they reveal the more personal motivations behind the activities of the first generations of wage-earning women in modern Korea.

For many young women, industrial employment altered their positions within their families during a pivotal stage in their lives. Working away from home, women became more materially independent of family control. But the effects of mill life were also psychological. Factory work brought on drastic transformations of habits and surroundings; these changes inspired different modes of identification. For the first time in their lives, industrial labor compelled women to view their social positions from reference points outside of the family and community. Discovering their standings in larger settings, such as the factory, might have been the first of numerous related disappointments for women such as Kang Pokchŏm. For others, including Kim Chŏngnam and Kim Chŏngmin, their labor bolstered self-confidence, and although they endured injustices as young women of lower-class backgrounds, they overcame gender- and class-specific obstacles later in their lives.

Commencement of work was a rite of passage and many entered the factory as children but left matured, having coped with the productive responsibilities of adults and men. Individually, wage work prompted revisions in life plans, priorities, and expectations. Collectively, employment engendered new modes of self-identification that integrated factory girls' social roles as women and workers. Mastering semi-skilled and technical duties at young ages roused self-assurance and encouraged later professional pursuits. Experiencing cultures abroad also brought, for many, diasporic experiences, identities, and values. Although the life choices of some might suggest radical shifts in family expectations, the working women of wartime Korea often fulfilled their parents' long-range goals of social mobility. As independently as former factory women might have lived, they were still affected by their parents, family, kin, work relations, and other communities of interests. Material and emotional autonomy, just like codependence and independence, were not possible without the cooperative interaction between individual time, family time, work time, and historical time at varying points in these women's lives. The needs and wishes of the collective groups surrounding the individual were thus inseparable from personal ambitions.

Industrial labor was a transitional experience in the trajectory of life stages. For women workers of late colonial Korea, the factory served as a halfway point between childhood and adulthood. Wartime factory girls were also transitional figures in their family histories. Because the reasons for sending daughters to work in early twentieth-century Korea relied on the needs of family economies, most factory girls grew up poor. Still, wage work extended beyond its material function and determined women's views of themselves, their life plans, and their aspirations. Testimonies imply that entering the workforce led rural girls to consider the possibilities of future careers for the first time. The youth of late colonial Korea labored in the factories for only a few years, but most of these women continued to work for wages throughout their lives. In many cases, women improved their financial positions through wage work and achieved educated, respected statuses, if not for themselves, for their children.

As scholars of economic and women's history have revealed, industrialization did not occur in a revolutionary manner but was characterized by sporadic currents of economic development. The occupational transitions from farmer to factory worker to professional rarely occurred within a single generation. Therefore, the Korean women who entered wage work during the Pacific War were members of this transitional generation that shaped the social progressions of the latter half of the twentieth century. The experiences of working women in twentieth-century Korea indicate that modernization was and remains a process that extends beyond periodic boundaries. Activists have depicted their lives to exemplify nationalist, feminist, and working-class interests, but a critical evaluation of the working women of colonial Korea demonstrates that they held to identities and alliances that altered with time and circumstance. Social mobility might have meant a break from tradition, but in practice, the attainment of knowledge and wealth represented the selective continuation of the conventions and values of parents and grandparents. Indeed, female factory workers of the late colonial era embraced progressive ideas concerning their position and value in economy and society. But, just as consequentially, they simultaneously held to their conservative roots as transitional figures in personal, familial, political, and historical continua.

# The Legacies of Colonial Working Women

On November 13, 1970, at the P'yŏnghwa market in Seoul, a garment worker poured gasoline over his body and set himself on fire. Holding a copy of the Labor Standards Laws, Chŏn T'aeil's suicide protested the blatant violations of these laws by employers. While drastically different in their outcomes, Chŏn's symbolism mirrored that of previous generations of workers, such as the activism of Kang Churyŏng who, to draw attention to the inadequate conditions of rubber shoemakers, climbed on the rooftop of the Ŭlmil Pavilion in central Pyongyang on the morning of May 29, 1931, refusing to come down.[1] Like their colonial predecessors, industrial laborers in late twentieth-century South Korea confronted arduous hours of work. But, whereas in the 1930s, the longest shift recorded was fourteen hours a day, in the 1970s, children and women labored almost continuously for up to sixteen hours a day in Korean-owned manufacturing mills. The thriftiest of employers, to cut costs, halved the floors of their workshops. Literally reducing overhead, women and children were forced to toil in rooms 1.6

meters high.[2] Impoverished labor conditions in post-Liberation South Korea eclipsed Japanese colonial models. Intended to draw attention to the severity of these abuses, Chŏn's martyrdom served as a catalyst for workers' activism throughout the 1970s.[3]

Suppressed in the last years of the colonial era, civil protests as well as labor and peasant organizations surged after August 15, 1945. Despite the emergence of separate states in the North and in the South by 1948, initial efforts for labor organization after Liberation were pan-Korean. Union leaders from throughout the peninsula met in Seoul on November 5 and 6, 1945, inaugurating the left-wing Chosŏn National Council of Labor Unions (Chosŏn nodong chohap chŏn'guk p'yŏnghŭihoe) with a membership of half a million. As the United States Army Military Government in Korea (USAMGIK), instated in the South on September 8, 1945, planned to ban labor organization, the unions of the North reassembled into the North Korean Federation of Trade Unions (Puk Chosŏn chigŏp tongmaeng), becoming totally independent of the Chosŏn National Council, with its headquarters in Seoul, by May 1946. Notwithstanding official suppression, the Chosŏn National Council of Labor Unions successfully coordinated a series of strikes throughout the U.S.-occupied South in 1946. These included the shutdown of the Tongyang (formerly Tōyō) textile factory in Inch'ŏn in June, the general railway workers' strike in September, and the Taegu riots in October. Nevertheless, with the assistance of right-wing labor groups, most prominently the Korean National Labor Union Association (Taehan min'guk nodong chohap ch'ŏngyŏnmaeng), the military government banned assembly and arrested dozens of its top leaders, terminating the Chosŏn National Council by February 1947.[4]

Irrespective of their pan-peninsular intentions, initiatives for improved labor and gender relations evolved quite distinctly in the North and in the South after 1945. In the North, labor and social insurance laws, including the "eight-hour workday, equal pay for men and women and accident and health insurance," were adopted in June 1946. Similarly, the gender-equality law, passed by the North Korean People's Provisional Committee in 1946, gave women parallel rights to political participation, economic opportunity, education, and "freedom of choice in marriage and divorce."[5] In the immediate aftermath of Liberation, corresponding developments occurred in the South. People's committees (*inmin wiwŏnhoe*) and factories in self-governing councils (*chach'i kwalli wiwŏnhoe*) sprouted and spread throughout the population. These committees led to the formation of the Korean Peoples' Republic (Chosŏn inmin konghwaguk) on September 6, 1945, and its adoption of a twenty-seven-point platform, which included universal suffrage, the prohibition of child labor, and the institution of eight-hour workdays and minimum wages. Nonetheless, the USAMGIK's dissolution of the Korean Peoples' Republic and the subsequent establishment of the Republic of Korea on August 15, 1948, impeded many political and practical movements for social and economic equality.

As in the North, formerly private, colonial enterprises in the South such as factories, mines, and construction projects were annexed by the state. The U.S. military government in the South, under Ordinance Number 33, took over the operations of over three thousand of the region's largest businesses and institutions immediately after occupation. Although nationalization revitalized these industries, many of which had deteriorated during the Pacific War, the Korean War devastated not only production but commercial infrastructure throughout the peninsula. The pledge of the U.S.-Korean Joint Economic Committee Concerning Economic Reconstruction and Financial Stabilization, or the Paek-Wood Agreement, concluded in May 1953, assured the South approximately one billion US dollars in aid, much of which was used for economic rehabilitation. The reconstruction policies of the 1950s encouraged the privatization of industries in the South. Colonial spinning and weaving mills were taken over by Korean businessmen. Pak Ch'anju, for instance, founded Chŏnnam Textiles in 1952 in place of the old Kanegafuchi factory in Kwangju. The Tōyō factory in Inch'ŏn was reestablished as Tongyang Textiles under the leadership of Sŏ Chŏngik in 1955.[6] Cotton textiles and other female labor-intensive projects, first endorsed as a vital part of the 1950s post-war import-substitution policy, continued to flourish in the 1960s and 1970s, as state-led export drives shifted the focus of production back to the manufacturing of consumer goods.[7]

In the decades following reconstruction, Korean women again entered the mills, making silk and cotton cloth, processed foods, rubber shoes, and machinery. Employing between two and three thousand workers, most of these factories resembled colonial enterprises in terms of their scale and strained labor management relations. Thus, following another phase of rapid development and repressive labor control, women workers rose to action once more. The strikes of Tongil Textile operatives, formerly the Tōyō Spinning and Weaving Company (Tōyō bōseki kabushiki kaisha), achieved notoriety by disabling production for almost three years, from 1976 to 1978.[8] Like women of the colonial Tōyō Company, the employees of the succeeding Tongil Corporation objected not only to the deficiency of their conditions but to the inadequacies of the labor practices and policies prevalent in South Korea.[9]

According to popular history, whereas colonial workers protested for anti-imperialist reasons, the strikes of the 1970s were stirred by working-class consciousness. Indeed, the labor movements of late twentieth-century South Korea finally pushed through comprehensive laws improving shop-floor settings and protecting workers' rights. Nevertheless, such advances would not have been possible without the efforts of preceding generations. Workers after Liberation not only drew upon the organizational designs of early twentieth-century activists but also acquired greater efficacy through public education introduced in the colonial era. Strengthened by training and supported by a populace familiar with the problems of modernization, the campaigns of the 1970s and 1980s led to the establishment of the eight-hour workday, minimum wages, and safety regulations on the shop floor. Despite the tendency to trace labor and civil reforms to post-Liberation efforts, the rebuilding of society, whether for class or for gender equality, began earlier and from below.

After a period of relative quiescence, Chŏn's dramatic death presaged a labor movement unparalleled since the 1930s. Not only did it ignite a resurgence of union and strike activities, it also led to the founding of an array of civil organizations that sought to implement the social reforms ignored for decades because of war and accelerated development. In the fifteen or so years following Chŏn's death, hundreds of labor, women's, and other civil associations were founded, among them the Society for the Survivors and the Bereaved Families of the Pacific War, which in 1973 established its headquarters at Yŏngsan. It was during this time that activists and intellectuals started to circulate revisionist exposés such as Kim Ilmyŏn's 1976 account of comfort women.[10] The civil renaissance that emerged in the 1970s was closely linked to colonial activism in not only its proposals but its participants. That many people who worked in the colonial era registered with the society from the 1970s on conveys the continuation and evolution of women's political consciousness and action in South Korea. The 1970s ushered in an age when the plight of women workers was no longer connected to imperialism or outside influence but to distinctly domestic and "Korean" customs of sex- and class-based discrimination. Notwithstanding the subjectivities of memory, by participating in the civil campaigns of the late twentieth century, former factory girls acknowledged their pasts as mediated by factors other than ethnic subalternity.

Social environments, often affected by nationality, class, and gender, configured individual consciousness, but not in uniform or timeless ways. An emphasis on social environments underscores the significance of spatiotemporal particularities, not to mention the personal and interpersonal idiosyncracies, involved in identity formation and maintenance. For colonial factory women, the social and economic transitions of the early twentieth century were unprecedented. Whereas in the 1910s, the number of mills in Korea was in the few hundreds, it multiplied in the 1920s and reached several thousand by the early 1930s. In 1910, less than 15 percent of the population occupied itself with nonagricultural work, but, by the 1940s, over one-third of the populace was engaged in manufacturing and commerce. Such figures, viewed in conjunction with the fact that approximately half of the agricultural populace also worked part-time for wages in the 1930s, clearly point to the expansion of wage work in early twentieth-century Korea. In the fifteen-year span between 1925 and 1940, the urban population on the peninsula rose by more than 300 percent. Around four million Koreans had migrated to Manchuria, Japan, the United States, and Russia by the end of the colonial era. Such dramatic and rapid historical transformations uprooted and reconfigured the social contexts of Korean life.

With the agricultural commercialization and industrial development coinciding with colonization, much of the peasant population left the countryside and moved to urban areas for work. Although some families migrated as a unit, unmarried young women were among the first to enter urbanized factory labor independently. Throughout the twentieth century, textile labor was seen as an extension of domestic service, but work inside the mill gates was anything but customary. While women in premodern Korea worked outside of their households, they rarely ventured outside of the domains of their community. Because familiarity, or expectation, affected perception, women might not have perceived gender inequalities within familial or communal settings. Nonetheless, as early Korean feminists promoted improved women's rights upon encountering the challenges of formal education, factory women perceived their "femininity" and social roles differently after the commencement of mechanized labor. The factory system was no more paternalistic than the tradition of service work, but injustices, when viewed in relation to supervisors rather than elders or house masters, became more conspicuous. Whereas scholarly work uplifted elite feminists from the bounds of womanhood by offering radical alternatives, factory work presented no such relief. Rather, their experiences made these informal confines of gender even tighter and thus more apparent. In confronting and overcoming the everyday abuses of paternalism, factory women in colonial Korea dealt practically with the problems posed by capitalist patriarchy.

Deconstructing the myths of women's and workers' collectivity suggests that women's political consciousness was personally experienced and conceived. In contrast to earlier descriptions of the nationalist and class consciousness of women, the modes of association that working women embraced did not relate exclusively to the nation or to class. Because one's sex frequently determined the kinds of jobs performed before and after marriage, working women were most likely conscious of gender inequalities, but their identities were distinct from those of elite feminists. With the expansion of wage work, women gained a new kind of consciousness connected to the tasks they performed and their monetary value. But their identifications with their labors were not permanent. Just as former factory women empathized with the rigors of industrial production, they appreciated the difficulties of motherhood and domestic labor they encountered later in life. Thus, the political consciousness of working women evolved through the myriad of tasks accomplished in everyday life. The collective groups that workers identified with were not the unified fronts of nationalists, socialists, or feminists but those that they interacted with on a daily basis. Although their more familiar attachments have been overlooked in many studies of colonial factory workers, most Korean women were still affected by the values they acquired from youth. Rather than adhering to permanent alliances, the changing identities of working women allowed for the transformation and growth of individuals, families, and communities.

## The Identities of Factory Women

Although early twentieth-century scholarship relied on constructions of coherent selves, social identity has been shown to be a "front," as that part of the individual's performance "which regularly functions in a general and fixed fashion to define the situation for those who observe the performance." Similarly, proponents of identity theory maintain that people's impressions of individualism are internalized role designations "carrying the shared meanings and behavioral expectations associated with roles and group memberships." Like the dramatization of a play, performance constitutes the coherence of an identity within the receiving social framework. Structure is constructed through enactment, and thus performance is an instrument of individual and social evolution. The creation and maintenance of people's identities takes many forms: stories, rituals, and dramatizing claims that render visible the actual and the desired truths about individuals. These enactments are not merely for self-recognition but are critical contributions to evolving processes of self-definition.[11]

Social roles, or performed identities, are also reliant on habits, which can determine an individual's commitment to an identity. A repeated performance, by reducing the consciousness of accustomed stimuli, allows for abstract thought and creativity. Nowhere is the productive potential of habit more evident than in the arena of work. Industrialization and urbanization, by forcing people to leave local communities for wages, alienates people from their work. But modernization also coincides with heightened specialization that caters to the refinement of skill. In work, the individual seeks not only financial returns but gratification of the particular pattern of "instinctual wishes and desires that comprised their unique character or temperament."[12]

Modern management procedures, including those used in colonial Korea, increasingly treated workers as individuals in search of responsibility, achievement, and meaning. Individuals were not to be emancipated "*from* work, but to be fulfilled *in* work," now construed as activities through which humans produced, discovered, and experienced "selves." Although the self-actualizing characteristics of labor have been associated with wage work, feminist theory offers alternative perspectives. If, departing from conceptions of work as a means to an end, it is conceived more broadly as the output of energy, all forms of human effort can be perceived as contributing to the formation and maintenance of identities. Whether efforts were expressed through wage labor, domestic management, childbearing and child rearing, or time spent with family members and friends, working habits were central to configuring selves.[13]

Among the theories that expose the correlations between identity and agency is the self-concept, which helps explain motivation. According to Viktor Gecas, self-concepts rely on self-esteem, self-efficacy, and authenticity. Self-esteem is reinforced by performed acts that are designed to increase sentiments such as pride while decreasing those such as shame. Closely related to self-esteem is self-efficacy, which refers to the "motivation to perceive oneself as a causal agent in the environment." Authenticity confirms the validity of these beliefs. The combined objectives of esteem, efficiency, and authenticity inspire the creation and maintenance of certain identities and behaviors in individuals while restraining other identities and behaviors.[14]

Reexamining the activities of colonial women workers with theories of identity in mind indicates that undergoing wage work for many factory women nurtured their consciousness of skill. Although labor demonstrations in colonial Korea were incited by unfair treatment, it is important to remember they were simultaneously forged by sentiments of professional entitlement. During the first decade and a half of industrialization, from the late 1910s to the mid-1930s, the number of strikes and boycotts rose dramatically, conveying that workers expressed their internalized values through political action. The injustices of Japanese wage scales and labor policies have been perceived as the motivators of strikes, but such interpretations gloss over the specialized ways in which protests arose. Colonial capitalism established certain wage standards, such as the well-known fact that Koreans received half the wages of the Japanese, and women, half the pay of men. The system of inequality supported by Japanese rule formed common tensions for Koreans and most likely exacerbated the financial problems of lower-class Koreans. Still, colonial demonstrations were not motivated by unequal structures; they exploded when employers disturbed this "stasis of inequality" with further pay cuts or mistreatment.[15]

Labor activism in colonial Korea was presumably inspired by the desire for national liberation. But the fact that demonstrations occurred predominantly in times of decreased economic growth clearly shows they were more likely prompted by material vulnerabilities. Whereas in 1921, there were thirty-six reported strikes in Korea, by 1931, this number rose to 201. Especially in the few years before and after the Depression, a boycott broke out nearly every day in metropolitan centers such as Seoul, Pusan, and Pyongyang. Over 2,100 labor demonstrations, involving over 100,000 men and more than 60,000 women, erupted between 1920 and 1938, when the first modernization efforts of twentieth-century Korea emerged and expanded. Although scholars have downplayed the importance of many colonial strikes because they were dispersed, a detailed analysis of women's organizations suggests that factory and industry-specific uprisings represented the broad spectrum of interests among different workers in the early twentieth century. Rather than the unified front of the nation or the working class, explicit problems, such as drastic cuts in wages and cases of mistreatment, provoked most labor disputes. Protests emerged in the 1920s as the first industrial workers took action against the shortcomings of early development and high inflation rates in the years following WWI. The sharp rise of workers' boycotts in the years surrounding the Depression reflected the need for immediate economic relief rather than ideological alliance, as wages hit record lows. The vast majority of colonial workers demonstrated for specific reasons concerning compensation and treatment, often unrelated to the greater theme of national independence.

While the activities of men in heavy industries are better known, protests by working women were just as pronounced. Between 1929 and 1934, more than one hundred large-scale demonstrations surfaced in female labor-intensive enterprises such as silk-reeling, cotton spinning and weaving, and rubber shoemaking factories. Because filatures generally hired fewer employees than cotton spinning and weaving factories, the boycotts of silk-reeling women were small in scale. Nevertheless, women employed in cotton textile projects organized large-scale demonstrations. For example, one of the largest boycotts of the colonial era, the 1930 strike by workers of the Chōsen Spinning and Weaving Company in Pusan, with over 2,200 participants, was composed mostly of women. Female workers also executed citywide protests such as the general strike of Pyongyang rubber workers in 1930. Although wages were the primary concern, far-reaching objectives such as financial compensation for work-related injuries and demands for weekly holidays were also voiced. Working women raised gender-specific issues, submitting plans for maternity leave and childcare facilities. Such proposals show that though women called for the immediate relief in wages, they also proposed long-ranging reforms. The fact that the workers of the 1930s called for the revision of corporate practices that continue today, such as night work and production-based wages, evidences the levels of sophistication that women exhibited in early twentieth-century Korea.

Just as male employees in the colonial era did not join forces with all other Koreans, or even all workers, women were far from unified. The demonstrations of female factory operatives were not exemplary of a united front of working women. Although workers in certain sectors, such as those in knitting and rubber shoemaking, were able to assemble for general strikes, the vast majority of boycotts in colonial Korea were factory and case specific. Even general strikes did not mean greater solidarity for all workers. As exemplified by the reports of the Pyongyang amalgamated rubber workers' strike in 1930, female shoemakers strongly objected to the recruitment of young unmarried women, placing their interests before those of a competing group of women. Actions and identities evolved concurrently as workers actively resisted those who challenged their own assessments of their authenticity, expertise, and value. Although factory women held strong convictions concerning their labors, their collective activities were also extensions of their daily associations.

## The Agencies of Factory Women

If eighteenth-century philosophical traditions inflated individual ability, twentieth-century theories deflated personal agency vis-à-vis the power of surrounding social structures. Much of this scholarship reveals fundamental flaws in liberal premises, but simultaneously discourages attempts to understand people's tendencies to conceive of individual influence.[16] Because individuals and social structures share power, agency is neither unlimited nor necessarily causal. Moreover, research on the unconscious suggests that many human behaviors are not overtly motivated. Direct causal impact is further debilitated by the centrifugal forces of social life, which indicates that actions are not just consciously and unconsciously driven, but have intended and unintended consequences. Although intentionality is frequently implicated in understandings of agency, according to Anthony Giddens, agency "refers not to the intentions people have in doing things but to their capability of doing those things in the first place."[17] Rather than distinct, states of independence and dependence can be seen as fluctuating positions along a spectrum of power that people exercise in spatiotemporally and socially diverse contexts. Movements between the poles of individual and sociopolitical authority, independence, and dependence are, rather than dichotomous, discursive.[18] That agency is not always intentional, however, does not mean that individual influence is not determinable.

Factory women gained leverage through unified action, but their agency was most visible in their everyday associations within the mills. Rather than membership in the working class, the consciousnesses of factory women were shaped by specific efforts, experiences, interests, and values. Women endured, resisted, and negotiated the factory system within smaller groups based on affinities of birthplace (*kohyang*), age, education, religion, and personal background. Rather than by common membership in a proletarian class, workers were motivated by common roots in the countryside, common circumstances in the factory, "common dialect and customs, common problems in times of war and economic crisis."[19] Although times of distress fostered sentiments of mutuality among factory women, their activities and affiliations within the structures of everyday life also reinforced these alliances. Many readily adapted to their tasks as manufacturers, but others held firmly to their rural identities. Throughout their lives, working women in Korea vacillated between alliances based on ethnicity, class, gender, and age in various circumstances, but the collectives that remained closest to individuals were communities and families. Just as factory women ran away together, they endured imprisonment, hunger, physical pain, and unyielding labor with friends and associates, who helped each other continue their contribution to the family economy.

Everyday life, however, was not just reliant on immediate events but on gradual historical developments. From the late 1910s to the mid-1930s, a generation of Koreans confronted the trials of early industrialization following the First World War, agricultural reform, economic recession, and the Great Depression. No preceding event, however, transformed the lives of ordinary people to the extent of the Second World War. Although the war divided, confined, and inhibited people from pursuing everyday activities, paradoxically it also prompted "enormous changes in the relationships among the groups and classes" that constituted society.[20] After twenty-seven years of colonial rule, Korea was propelled into an unprecedented international war as part the Japanese empire. Launched in 1937, Japan's war against China and the ensuing war against the Allies provided the first occasion for select lower-class women to transcend traditional class- and gender-based boundaries.

These larger historic transitions, although offering opportunities for some women, still had undesirable effects. The testimonies of three former employees of textile factories in Yŏngdŭngp'o and Kwangju, established in the 1920s, attest to the prevalence of rudimentary labor conditions even in the late 1930s and early 1940s. Two former workers of the Kanegafuchi factory in Kwangju, Yi Chungnye and Yi Chaeyun, depicted the kinds of shop-floor settings that would have been familiar to textile laborers in the 1920s: twelve-hour shifts, two days of rest per month, draconian supervision, frequent accidents, and death.[21] Despite some improvements in work and living standards by the late colonial era, the recollections of the former operatives of spinning and weaving mills confirm that corporate improprieties continued. Kang Pokchŏm's account of her four years of factory life at the Tōyō Spinning and Weaving plant in Yŏngdŭngp'o echos published descriptions of the shop-floor environments of the 1920s and 1930s. Just as significantly, it vividly details how some factory women endured these terms of labor through their unique methods of accommodation, collaboration, and resistance. Working in mills established in Korea for nearly twenty years, Kang Pokchŏm, Yi Chaeyun, and Yi Chungnye shared the same difficulties of the former generation of workers.

Still, the ever increasing demand for human resources brought on by the Pacific War allowed others to rise to positions of skill. Recruited directly from the countryside, Yi and Kang were placed in unskilled positions, whereas female students with schooling became regular workers (*honshokkō*) in heavy industries. The testimonies of Kim Chŏngnam and Kim Chŏngmin, who worked at the Fujikoshi Steel factory in Toyama, clearly illustrate that the war economy opened a window of opportunity, however slim, for ambitious lower-class women. Joining the women's volunteer corps through public school recruitment, they rose to skilled positions and continued their education after Liberation. Women like these, who became professionals after the war, did not confine their consciousness to the bounds of the working class but, rather, saw their factory work as the first of many jobs they encountered throughout their lives.

Some of the stories of wartime workers might suggest that the employees of textile companies in Korea endured severe labor conditions, which contrasted with those of factories in contemporary Japan. The descriptions of Han Kiyŏng, who worked in the later-established Dai Nippon Spinning and Weaving plant in Yŏngdŭngp'o, however, convey that gradual improvements in management systems were discernable even within the Korean textile industry. In contrast to the recollections of Yi and Kang who endured twelve-hour shifts, Han reported having eight to nine hours of work a day, four rest days a month, and cooperative relations with superiors. Those "forced to labor" in colonial mills were seen as the worst exploited workers, but the accounts of former factory women show that the mill also served as the breeding ground for their political consciousness and even productive confidence. Whereas some adapted quickly, others never adjusted to the pace of the factory. Women recruited directly from the countryside expressed the greatest difficulty in adapting to industrial life, but even semi-skilled workers in sophisticated enterprises might have favored other jobs over mill work. As Pak Sun'gŭm, a former lathe worker at Fujikoshi, maintained, whether in Toyama or Seoul, she always preferred "country living" in North Chŏlla province.[22]

## Communities, Families, and Work Throughout Life Courses

Nineteenth-century socialists envisaged that the alienation of modernity would bind people together for revolution. Recent feminist and postcolonial critiques, however, indicate that the rationales for resistance and revolution are not based solely on socioeconomic class. Recognizing that colonial conditions cannot be explained by capitalist logic alone while adhering to unilateral interpretations of domination, postcolonial scholars have referred to those disenfranchised by patriarchy, coloniality, and capital as the subaltern. Despite the descriptive value of subalternity, by retaining the logic of defensive power, many have portrayed colonials, workers, and women as reactive rather than proactive people. Furthermore, the amalgamation of various social groups, such as workers, women, and postcolonial populations, into the category of the subaltern unwittingly diminishes the differences between and within this disenfranchised mass. Exploring how moralization mediates between individual and collective ideations and actions provides a way of understanding agency beyond the subaltern perspective, as not a categorical, but a contextual, force.

Few human inclinations better convey the connection between cognition and action, and the individual's innate potential for power, than the pervasiveness of moral positions. According to G. H. Mead, "enlarging the self" through moral development, by improving role-taking abilities, or by "learning how others perceive, interpret and respond to the world" is critical to identity formation. Moral or ethical positions come to life in environments of conflict or discontent, when habits alone are insufficient guides to activity. By projecting certain courses of action within specified situations, these stances rely not only on individuals' imaginings of social responses but also on individual commitments to, and expectations of, others' adherence to ascribed behaviors. Thus, factory women's acts of resistance can be seen as responses to violations of communal norms. Strategies of protest thus catered to the moral sentiments of opposing parties. Positions of victimhood, for instance, were performed to audiences "to authenticate identities" and to mitigate negative responses. Resistance, an appeal to structural integrity, required the individual's belief in his or her ability to alter surrounding social contexts, thereby magnifying the meaning of the individual's existence.[23]

While workers' demonstrations expressed moral, or by extension political, positions, the mutability of resistance movements explains the inconstancy of political constituency. As Judith Howard asserts, moral commitments could not operate as incentives for activism if "individual values, emotions and identities" were not critical components of people's decisions to join movements. Nonetheless, in most spheres of life and in most types of activity, "the scope of control is limited to the immediate contexts of action or interaction," suggesting that individual agency was exercised not only in macro realms but in micro communities as well. Because belief in efficacy, or authenticity, was central for identity maintenance, people did not attempt to reform macro social structures repeatedly when previous attempts were unsuccessful. Instead, they focused their energy and influence in smaller communities, in the neighborhood, at work, at school, and at home.[24] More often, the immediacy of relations in the factory, the community, and the home took precedence over the resolution of overarching moral problems.

If individuals were not loyal constituents of political groups, to whom were they bound? People were most devoted to those ties that sustained the self because the salience of an identity was a direct result of commitments to specific networks of social relationships. But identities were not permanent; because individuals asserted more agency in partnerships, friendships, and familial relations, performance relied just as much on the evolving dynamics of these relationships as on designated roles. Both emotively and materially, family relations and other ongoing social affiliations were critical to the creation and recreation of identities and the proper functioning of the self.

The reasons for sending daughters to work in early twentieth-century Korea depended on family circumstances. With the expansion of the wage economy, tasks and obligations were given monetary meanings, but the reasons for labor remained unchanged—people worked for their families.[25] Although many Korean families remained in the lower classes throughout the twentieth century, few sources indicate the generational continuance of industrial labor. Thus, although factory women might not have attained social or professional mobility within their lifetimes, their children often fulfilled their long-range family goals. Like jobs, careers were also "shaped by the age configurations and economic situation of the family."[26] Individual career decisions relied on a combination of personal attributes, choices, and skills, affected by the timing of the first job, age, experience, and the availability of employment. Despite the difficulties of manufacturing, testimonies indicate that factory work extended beyond its defensive function and stimulated women's visions of their life plans and aspirations. In encountering their first jobs, young rural women began considering the possibilities of future careers and future selves.

Just as historical milieus placed women workers in distinct contexts, so did family circumstances. For better or for worse, young women of the 1920s, 1940s, and 1970s learned life's harsher realities by entering the factory. Colonial women, especially those who started wage work during the war, departed from the customs of their families and communities throughout their life courses but fulfilled the aims of their parents by providing for their children. Whether women obtained respected positions or not, they did not hesitate to discuss setbacks and hardships inasmuch as the obstacles they overcame highlighted the extent of their achievements and strengths. Just as they did not denounce all aspects of their femininity, working women did not see their struggles as exclusively mediated by class or nation. In explaining their consciousness, informants described their multiple identities and understandings as colonial subjects, women, workers, mothers, professionals, and individuals. In encountering wage work, many remained in the public sphere throughout their adult lives and, in so doing, acquired greater economic and emotional independence.

*Summation*

In *The German Ideology*, Karl Marx combined his interpretation of history with his theory of consciousness. Reversing the procedures of "idealists" who took consciousness as a priori, Marx proposed that the production of ideas and identities was inseparable from human activity and discourse:

> In direct contrast to German philosophy [that] descends from heaven to earth, here we ascend from earth to heaven. We set out from real, active men [and women], and on the basis of their real life-process we demonstrate the development of the ideological reflexes and echoes of this life process. . . . Life is not determined by consciousness, but consciousness by life.[27]

That consciousness was socially construed had political implications for the materially dispossessed as historical progressions altered social conventions and ideals. Class consciousness, as Marx conceived, would rise simultaneously with the changes in material reality ushered in by capitalist development. By connecting mind to matter, Marx departed from rationalist traditions, but in so doing reinforced the constancy of consciousness after modernization. Although spurred by socialism, nineteenth-century nationalists, suffragists, and activists ironically also took consciousness as a priori, suggesting that sociological traits, such as ethnicity, sex, and class, played central roles in the formation of individual identities, alliances, beliefs, and behaviors. Nineteenth-century proponents of "consciousness-raising" campaigns failed to realize that apart from gender, class, or nation, the more specific fluctuations and interactions of history and locality formed and transformed individual beliefs and actions.

Despite the rhetorical unity of the "working class," historical time affected the conditions and meanings of women's work, making factory employment far from constant. Arising in the 1920s and 1930s, the first generation of workers in twentieth-century Korea encountered the difficulties of early industrialization and crude shop-floor settings as well as rudimentary management relations. Although these young women suffered from some of the worst abuses of capitalism, they were among the first to voice their collective proposals for improvement and aims for the future. Nevertheless, the characteristics of women's work altered by the late 1930s, as some men and women came "to the age of work when there [was] no work," and others, during war.[28] Entering the labor market amid a period of shortage, women who commenced work in the late 1930s and early 1940s encountered a greater range of employment opportunities. The configuration and dynamics of women workers in the late colonial era confirms that the second generation of wage-earning women in Korea had a wider array of professional prospects, including those in heavy industries.

Just as mechanized development in the 1920s and 1930s led colonial Korean women to the mills and other forms of wage work, modernization drives after Liberation again transformed South Korean women's lives and labors. Similarly, post-Liberation civil and labor initiatives drew upon earlier forms of activism. Employed in factories of comparable scale and organization, the petitions of colonial workers such as the reduction of work hours and workers' compensation were again raised in the 1970s. Gender-specific provisions including maternity leave and equal pay were similarly proposed by the later generation of working women. Although the movements of the 1970s and 1980s resulted in the adoption of more equitable labor laws in Korea, some of the practices introduced in the colonial era continue. Rural youth are still sometimes recruited on middle school grounds, and the employment of minors fourteen and fifteen years of age persists. Night work, albeit with higher pay, remains and factories continue to operate machinery twenty-four hours daily. Some women still reside in dormitories that lock the gates each night.[29]

Apart from the pace of development, location was particularly consequential for South Korean workers. South Korea's geopolitical position as a
bulwark against communism in East Asia not only stimulated the economy
but made fundamental labor rights ultimately difficult to deny. Locality also
worked in reverse as regional and local customs might have supported the
perpetuation of gender- and age-biased employment practices grounded in
Confucian paternalism. Political positions involving nations and regions,
however, were probably not as instrumental to the worker as personal ones.
It is important to remember that the localities encompassing co-workers,
neighbors, friends, and family most closely defined individual activities,
identities, and values.

Rather than in overarching categories, the identities of working women
were anchored in their social relations. Although human cognitions, actions,
and reflections were contextually contingent, they were also adaptable,
which allowed for individuals to monitor the continuities of and modifica-
tions in their behaviors and identities. Born into poverty, prodded into the
factory, and forced into military prostitution, Kim Ŭnnye encountered
harsh relationships and harsher realities from an early age. Despite years of
tribulation, the fact that in her latter years she referred to her only daughter
and her grandchildren with confidence signifies the protean nature of
selves, communities, and possibilities. Through reflexive monitoring, individuals were influenced by and, in turn, affected the social surroundings
they encountered in the *durée* of everyday life.

Although frequently associated with nationalist, working-class, and feminist consciousness, former factory laborers did not perceive their successes
or setbacks as unilaterally determined by class, nation, or patriarchal traditions. Mill work and wage work in general brought new tasks, responsibilities, and experiences that for some exposed the social meanings of
lower-class Korean womanhood. While some working women in colonial
Korea were inspired by nationalism, ethnic homogeneity did not unite all
Koreans in everyday life. An ironic account illustrates this point. Kim Ŭnnye, a former operative in a hosiery factory, tobacco roller, and comfort
woman, served as a Korean representative for a 1999 United Nations hearing on the comfort woman issue in Geneva. Although she was a member of
the delegation, she complained that the Korean academics who took her
there did not treat her as an equal. She seemed slighted by their neglect, em-

phasizing that she only saw them a few times. Traveling to Europe for the first time at the age of seventy-four, her closest companion during the journey was another former comfort woman who was Japanese.[30]

The array of responses to modernization and wartime mobilization sheds light on how women's consciousness—particularly those identities related to work and, by extension, class—were not only varied but malleable throughout the life course. Rather than regretting their days at the factory, many women, including the workers of Fujikoshi, recalled with pride their wartime contribution. Two former factory girls, Kim Chŏngnam and Kim Chŏngmin, received college degrees and rose to respected positions in their professions. Inspired by her first trip abroad to Toyama, Japan at the age of twelve, Kim Chŏngnam continued her education after the war and eventually became a teacher. Similarly, Kim Chŏngmin used her earnings from the Fujikoshi factory to achieve a degree in Chinese medicine. She established a flourishing practice, married, and bore four daughters, divorcing her husband after "twenty-odd years." She eagerly added that he was not left indigent, but as a conciliatory gesture she "gave him a house." Kim Chŏngmin considered it an insignificant price to pay for the recovery of her freedom.[31] Some nationalists might argue that these women were pro-Japanese collaborators, indoctrinated to collude with their oppressors. Nonetheless, insofar as these young women's accommodation aided their acculturation with wage work, their wartime experiences might have eased their transition to modern life after the war.

Still other former factory women, including Yi Chungnye and Kang Pokchŏm, remembered more drawbacks than advantages to changing times and circumstances. Yi lost her right forearm as an apprentice at the Chongyŏn (Kanegafuchi) Textile Factory in Kwangju. Kang not only suffered from physical injury but lost a chance for a married life because of a crippling in-factory beating. Although they lamented their misfortunes, they nevertheless asserted confidence in their ability to conquer life's challenges. Yi maintained that her Buddhist faith provided an outlet for her to "do whatever she wants, like a man."[32] Kang, on the other hand, retraced a difficult life as she lived and worked independently. She did not express any pride, but observers would agree that her ability to live with exclusive self-reliance for over sixty years is a magnificent feat in itself.[33]

In conclusion, it seems appropriate to recite Patrick Joyce's words that readers may find a certain lack of conflict in this book, depending on "what one means by the term." Like Joyce, I've employed a broad definition of conflict, not only as political but as personal challenges whose outcomes are often "heterogeneous and ambiguous."[34] Throughout this work, I have proposed that conflict not only altered external structures but threatened and reshaped people's affiliations and identities. Repeatedly cited among informants' motivations for joining the corps was the desire to take part in a project larger than themselves or their immediate surroundings. The same sense of political consciousness that prompted these women to join the corps influenced their alliances, friendships, and political actions after the war. Throughout the 1970s, 1980s, and 1990s, hundreds of thousands of Koreans who had been mobilized for the war registered with civil organizations such as the Society for the Survivors and the Bereaved Families of the Pacific War. These activists included women such as Yi and Kang, who in seeking personal redress addressed the larger political injustices of the past.[35] Even in their twilight years, upon seeing each other again for their interviews, they reminisced and plotted resistance, as they had in their youth. It was as though their reunion sparked mutual memories that gave new life to old ambitions. As we sat down for dinner after a long day, Kang and Yi could not stop talking about their plans. To demand remuneration for their wartime work and injuries, they wanted to organize a sit-in before the presidential residence, *Ch'ŏngwadae* (Blue House), that Saturday. When asked why they chose the location, they attested that whereas the Korean government had compensated comfort women, those injured while working with the corps were left penniless. Interestingly, Kang and Yi directed their grievances at the Korean, not the Japanese, government. Throughout the evening, they argued over their plans. Kang wanted to meet after noon; Yi insisted that they arrive in the morning and protest all day. Talking, arguing, and laughing, they enthusiastically schemed, planned, and dreamed. These women might have lived with less official power in their everyday lives, but in their minds, ambitions, and friendships, they were agents of change.[36]

Although Korean women workers did not always hold idealized visions of feminist consciousness, they held convictions concerning their gender and its personal and political implications. The simple fact that they could step into men's shoes and "wear them comfortably" contradicted traditional notions of femininity and female deficiency.[37] Overcoming conventional limitations such as wage work and public activism during the formative years of their lives, working women in late colonial Korea continued to challenge patriarchal perspectives of female capacity throughout their adult lives. Women's public agency emerged and expanded throughout twentieth-century Korea. These advances were realized through not only ideology but the efforts of generations of working women. While some might have been motivated by nationalism, feminism, and socialism, the practical relations between teachers and students, mothers and daughters, and friends were far more influential, providing functional examples of alternative lifestyles and identities for the next generation. Former factory women recalled their collective experiences as students, workers, wives, and mothers, but these ties were not only temporally bound, they were also embraced and discarded circumstantially. Nevertheless, by confronting domestic realms and beyond, women workers gained new identities in relation to their evolving surroundings, and by acting with self-determination, they helped engender the political spheres around them. In view of the rise of gender, labor, and political consciousness as historical progressions, a question implied throughout the work might be raised: were the lives and sentiments of factory women as tragic as those depicted by ideologues? Although oppressed and victimized in popular memory, many working women of colonial Korea, in transcending both personal and political boundaries, saw themselves as pioneers.

# Notes

NOTES TO PREFACE

1. Joan W. Scott and Louise A. Tilly, *Women, Work and Family* (New York: Routledge, 1978), 2.

2. Edward Said, *Culture and Imperialism* (New York: Vintage Books, 1993), 3–14.


NOTES TO INTRODUCTION

1. Carter J. Eckert, *Offspring of Empire: The Koch'ang Kim's and the Colonial Origins of Korean Capitalism, 1876–1945* (Seattle: University of Washington Press, 1991); Dennis McNamara, *The Colonial Origins of Korean Enterprise, 1910–1945* (Cambridge: Cambridge University Press, 1990); Soon-Won Park, *Colonial Industrialization and Labor in Korea: The Onoda Cement Factory* (Cambridge, MA: Harvard University Press, 1999); Gi-Wook Shin and Michael Robinson, eds., *Colonial Modernity in Korea* (Cambridge, MA: Harvard University Press, 1999); An P'yŏngjik and Hori Kazuō, "Singminji kongŏphwa ŭi yŏksajŏk chogŏn kwa kŭ sŏnggyŏk" [The historical conditions and character of colonial industrialization in Korea], in An P'yŏngjik and Nakamura Tetsu, eds., *Kŭndae Chosŏn kongŏphwa ŭi yŏn'gu* [A study of industrialization in modern Korea] (Seoul: Ilchogak, 1993), 1–47; Kim Kyŏngil, *Ilcheha nodong undongsa* [History of labor movements under Japanese rule] (Seoul: Ch'angjakkwa pip'yŏngsa, 1992).

2. Because the wŏn was equal to the yen in value, the two were used interchangeably in primary sources. Chŏn, or sen, was 1/100 of a wŏn, or yen. In the late 1910s and early 1920s, when the first modern factories were built in Korea, one yen was equivalent to 0.40 contemporary US dollars. *Chosŏn chungang ilbo*, July 2, 1936.

3. Antonio Gramsci, *Prison Notebooks*, vols. 1 and 2 (New York: Columbia University Press, 1991); Michel Foucault, *Language, Counter-Memory, Practice: Selected Essays and Interviews*, Donald Bouchard and Sherry Simmons, trans. (Ithaca, NY: Cornell University Press, 1977), 206; Gilles Deleuze, *The Fold: Leibniz and the*

*Baroque*, Tom Conley, trans. (Minneapolis: University of Minnesota Press, 1992); Gilles Deleuze and Félix Guattari, *Anti-Oedipus: Capitalism and Schizophrenia* (Minneapolis: University of Minnesota Press, 1983); Gayatri Chakravorty Spivak, *A Critique of Postcolonial Reason: Toward A History of a Vanishing Present* (Cambridge, MA: Harvard University Press, 1999), 259; "Can the Subaltern Speak?" in Cary Nelson and Larry Grossberg, eds., *Marxism and the Interpretation of Culture* (Chicago: University of Illinois Press, 1988), 271–313.

4. Chŏk Ku (pseud.), "Yŏjikkong woesap'yŏn" [Women workers], *Kaebyŏk* 18:4 (April 1926), 110–11.

5. In the two Koreas established in 1948, as in other former colonies, the historical analysis of movements for social equality, including women's rights, have been interpreted in national terms. Nevertheless, studies that link feminism exclusively to nationalism or state socialism can offer much valuable and descriptive information. Jacques Derrida, "Women in the Beehive: A Seminar with Jacques Derrida," Transcript from the Pembroke Center for Teaching and Research Seminar with Derrida, *Subjects/Objects* (Spring 1984), 17.

6. An P'yŏn'gu and To Chinsun, eds., *Pukhanŭi Han'guksa insik* [Korean historiography in North Korea], vols. 1–2 (Seoul: Hangil, 1990).

7. Despite the merits of nationalist historiography in the two Koreas and elsewhere, a general overemphasis on attributing women's and workers' activities to nationalism has undermined the examination of other impulses for political activism in colonial Korea.

8. Although women in premodern societies most likely held protofeminist convictions, modernists hold that campaigns for political suffrage in the nineteenth and twentieth centuries engendered the birth of modern feminism. Founded by bourgeois Protestants, formative efforts toward gender-specific enfranchisement sought to expand the boundaries of the woman's sphere but ultimately embraced the "female identity that the cult of domesticity celebrated." Nonetheless, the efforts of conservative reformers laid the foundations for later feminists in their drive to promote an ideological shift from passive to active domestic influence. Barbara Welter, "The Cult of True Womanhood: 1812–1860," *American Quarterly* 28 (1966), 151–74; Yong-ock Park (Pak Yongok), *Hanguk yŏsŏng hangil undongsa yŏn'gu* [A study of the Korean women's anti-Japanese movement] (Seoul: Chisik sanŏpsa, 1996), 11; *Han'guk kŭndae yŏsŏngsa* [History of modern Korean women] (Seoul: Chŏngŭmsa, 1975), 13–31; Chung Yosup, *Han'guk yŏsŏng undong sa* [History of the Korean women's movement] (Seoul: Ilchogak, 1971); Lee Tae-Young (Yi Tae-yŏng), "Elevation of Women's Rights," *Korea Journal* 4:2 (February 1964), 5; Louise Yim (Im Yŏngsin), *My Forty Year Fight for Korea* (Seoul: International Culture Research Center, Chungang University, 1951); Pahk Induk, *September Monkey* (New York: Harper and Brothers, 1954).

9. Not all women who overcame the bounds of their gender sought to expand the rights of women more universally. Although female Christian associations were

frequently established abroad, marked religious activism began with the arrival of the "first [Methodist] representative of the Woman's Foreign Missionary Society to Korea," Mary F. Scranton, in June 1885. According to the records of the American Methodist representative in Masan, fifteen Bible women taught 145 rural peasants (*nongmin*) to read scriptures in *han'gŭl* in the year 1898. Whereas Mary Scranton zealously promoted the commissioning of native missionaries, Ella Appenzeller opposed such engagements, maintaining that "the women were still untrained 'raw heathen' and thus not qualified to teach others." Homer B. Hulbert, "Bible Women," *Korea Review* 6:4 (April 1906), 140–47; Mary F. Scranton, "Woman's Work in Korea," *The Korean Repository* 5:9 (September 1898), 313–18; "Among Women of City and Country," *The Korean Repository* 4:5 (May 1897), 294–97. For more on the Presbyterian mission's work on women, see Lillias H. Underwood, "Woman's Work in Korea," *The Korean Repository* 3:1 (January 1896), 62–65; *Fifteen Years Among the Topknots* (New York: American Tract Society, 1904); Chou Fang-Lan, "Bible Women and the Development of Education in the Korean Church," in Mark R. Mullins and Richard Fox Young, eds., *Perspectives on Christianity in Korea and Japan: The Gospel and Culture in East Asia* (Lewiston, NY: Edwin Miller Press, 1995), 29–30.

10. As she entitled the girls' school, Queen Min named the hospital the "Caring for and Saving Women's Hospital" (*P'ogun yŏgwan*). Esther Kim (1876–1910), also known as Esther Pak and Kim Chŏmdong, was the first Korean woman to receive a medical degree. Horace H. Underwood, *Modern Education in Korea* (New York: International Press, 1926); Kenneth M. Wells, "The Price of Legitimacy: Women and the Kŭnuhoe Movement, 1927–1931," in Gi-Wook Shin and Michael Robinson, eds., *Colonial Modernity in Korea* (Cambridge, MA: Harvard University Press, 1999), 211.

11. The Praise and Encouragement Society was associated not only with the Independence Club but another reformist organ, the All-People Society (*Manmin kongdonghoe*). When the two groups came into conflict with the monarchy and conservatives of the Imperial Society (*Hwangguk hyŏphoe*), female activists sought to join in the struggle, declaring that "it is an unprecedented honor as well as a right action as people to be imprisoned for the nation." Choi Sook-kyung (Ch'oe Suk-kyŏng), "Formation of the Women's Movement in Korea: from the Enlightenment Period to 1910," *Korea Journal* 25:1 (January 1985), 9; Lim Sun Hee (Yim Sun-hee), "Women and Education in Korea," *Korea Journal* 25: 1 (January 1985), 19; Vipan Chandra, "Sentiment and Ideology in the Nationalism of the Independence Club, 1896–1898," *Korean Studies* 10 (1986), 20–22; *Imperialism, Resistance, and Reform in Late Nineteenth-Century Korea* (Berkeley: University of California Press, 1988).

12. Italics mine. Cho Kyung-won, "Critical Examination of the Confucian View of Women's Education," in Jung Chang-young (Chŏng Chang-yŏng), ed., *Korean Social Sciences Journal* 151 (1990), 101; and Kim Yung-chung (Kim Yungch'ŏng), ed., trans., *Women of Korea: A History from Ancient Times to 1945* (Seoul: Ewha Woman's

University Press, Kwang Myŏng Printing Co., 1976), 223–25. For comparable initiatives in North America, see Daniel T. Rodgers, "In Search of Progressivism," *Reviews in American History* 100:12 (December 1982).

13. The students of private women's schools were, like factory women, generally housed in dormitories. Kim, *Women of Korea*, 260, 260 n7.

14. Kenneth M. Wells, *New God, New Nation: Protestants and Self-Reconstruction Nationalism in Korea, 1896–1937* (Honolulu: University of Hawaii Press, 1990), viii–xi; Yi Kyŏngsŏn, "Chosŏn yŏsŏng ege hosoham" [A petition for the women of Chosŏn], *Kaebyŏk* 22:1 (January 1930), 90–96; "Discussion on Women's Education," in *Yŏjagye* [Women's World], September 1919, as cited in Kim, *Women of Korea*, 259–60.

15. The *Aeguk puinhoe* was most reputed for its publication in the *Peking Daily News* on March 16, 1920, that called for international women's groups to assist in the cause for Korean women's and national liberation. The activities of the *Aeguk puinhoe* were recorded in Maria Pak's book. Maria Pak, *Kiddokkyo wa Han'guk yŏsŏng sasip nyŏnsa* [Forty-year history of Christianity and Korean women], in *Han'guk yŏsŏng munhwa nonch'ong* (Seoul: Ewha Woman's University Press, 1958).

16. Although many of the articles of the Kŭnuhoe journal, *Kŭnu*, focused on working women's problems, its contents frequently lacked detailed mention of either socialism or communism as offering a viable solution. For more, see Pak Hojin, "Yŏjikkong pangmungi" [Survey of women workers] *Kŭnu* (May 1929); Ch'u Ch'ŏn, "Yakhan yŏsŏng kwa nodong kyegŭp ŭi kiwŏn" [The frailty of women and the origin of the working class] *Sinyŏsŏng* 4:7 (July 1926), 36–37.

17. The platform of the Kŭnuhoe called for improvements to the conditions of women's work, proposing the abolition of wage discrimination, the termination of night-work for women and minors, as well as the institution of paid maternity leave. Although the composition of the Kŭnuhoe and its left-of-center political position allowed for the inclusion of such labor-related goals, its main agenda was the independence of the nation. The Kŭnuhoe manifesto concluded with grandiloquent maintenance that the liberation of the world rested with the liberation of women. "Charyo: Kŭnuhoe," [Kŭnuhoe manifesto] (Reprint) *Yŏsŏng* (March 1989); Wells, "The Price of Legitimacy," 206.

18. Na Hyesŏk and Kim Wŏnju both had unconventional personal lives. After the failure of her marriage, Kim Wŏnju lived as a Buddhist nun, whereas Na Hyesŏk abandoned her husband and lived independently. For samples of colonial women's writings, see Kang Kyŏngae, *In'gan munje* [Problems of humanity], Tonga ilbo serial publications (1934; reprint, Seoul: Ch'angjak kwa pip'yŏngsa, 1978); Kang Kyŏngae, Paek Sinae, and Kim Myŏngsun, *Haebang chŏn yŏru chakka sŏnjip* [Selected works of pre-liberation writers] (Seoul: Pŏmjosa, 1987); Yi Paeyong, "Ilche sigi yŏsŏng undong ŭi yŏn'gusŏng gwa kwaje" [The themes and findings of research on the women's movement during Japanese rule], *Han'guk saron* no. 26 (1992), 251–75.

19. As Frantz Fanon explains, a colonial, alienated from one's own country, lived in a state of "absolute depersonalization." Colonial nationalists posed many of the themes raised by late twentieth-century postcolonial critics who "intervene in those ideological discourses of modernity that attempt to give hegemonic 'normality' to the uneven development and the differential, often disadvantaged, histories of nations, races, communities, peoples." Frantz Fanon, *Toward the African Revolution* (Harmondsworth, UK: Pelican, 1967), 63; *The Wretched of the Earth* (New York: Grove, 1965); Homi K. Bhabha, *The Location of Culture* (London: Routledge, 1994), 171; *Nation and Narration* (London: Routledge, 1990).

20. Ch'oe Yŏnghŭi, Kim Sŏngsik, Kim Yunhwan, and Chŏng Hosŏp, eds., *Ilcheha ŭi minjok undongsa* [The nationalist movement under Japanese rule] (Seoul: Hyŏnŭmsa, 1982).

21. Im Chongguk, *Ilche ch'imnyak kwa ch'in-Ilp'a* [Japanese aggression and collaboration] (Seoul: Ch'ŏngsa ch'ulp'ansa, 1982).

22. I refer to colonization broadly as a process not limited to the imperial aggression of the nineteenth and twentieth centuries but including all forms of cross-cultural contact both preceding and succeeding the modern era. For more on the colonial and the modern, see Anthony Giddens, *The Consequences of Modernity* (Stanford, CA: Stanford University Press, 1990), 174; Mikhail M. Bakhtin, *The Dialogic Imagination: Four Essays*, Caryl Emerson and Michael Holquist, trans. (Austin: University of Texas Press, 1981), 358; Spivak, *A Critique of Postcolonial Reason*, 361.

23. Kang Tongjin, "Wŏnsan ch'ongp'aŏp ŭi taehan koch'al" [A study of the Wŏnsan general strike], in Yun P'yŏngsŏk, Sin Yongha, An P'yŏngjik, eds., *Han'guk kŭndae saron*, vol. 3 (Seoul: Chisik sanŏpsa, 1977), 237–68; Kobayashi Hideo, "1930 nendai zenhanki no Chōsen rōdō undō ni tsuite: Heijō gomu kōjō rōdōsha no zenesuto o chūshin ni shite" [The Korean labor movement of the early 1930s: A focus on the Pyongyang rubber workers' general strike], *Chōsenshi kenkyūkai ronbunshū*, vol. 6 (September 1969), 115; also reprinted in Namiki Makoto, et al., *1930 nyŏndae minjok haebang undong* [National liberation movements of the 1930s] (Seoul: Kŏrŭm sinsŏ, 1984), 241–69.

24. In premodern Korea, status was strictly defined. The social organization of traditional Korea was dependent upon a Confucian patrilineal descent system that ordained one's social, economic, and political position in life. Known as *yangban*, the elite of Chosŏn Korea constituted "not more than ten percent, of the total population; but, drawing on descent and heredity, it monopolized the political process," Confucian knowledge, and landed wealth. The class directly beneath the hereditary aristocracy was that of the good people (*yangin* or *sangmin*), usually termed *commoners*. The lowest class consisted of base people (*ch'ŏnmin*), "who consisted mostly of slaves but included those with base occupations, such as butchers, leatherworkers and shamans." With external pressure and dynastic decline, however, social statuses and alliances weakened. The status system of Chosŏn Korea, including the private ownership of slaves, was abrogated in 1894 with the enactment of the Kabo

Reforms. Martina Deuchler, *The Confucian Transformation of Korea: A Study of Society and Ideology* (Cambridge, MA: Harvard University Press, 1992), 12–13; Kyung Moon Hwang, *Beyond Birth: Social Status in the Emergence of Modern Korea* (Cambridge, MA: Harvard University Press, 2004).

25. Kobayashi examines the Pyongyang rubber workers' strike and places the demonstration in its industrial, regional, and historical context, but interestingly, "women workers" are hardly mentioned in his work. Although Kobayashi and others have assumed, or rather imposed, a misleading "masculinity" upon chemical workers, in reality, most workers in rubber factories were women. Kobayashi Hideo, "1930 nendai zenhanki," 115–17.

26. Kobayashi, "1930 nendai zenhanki," 115–22; Park, *Colonial Industrialization*, 123.

27. Leftist women's groups were allied with larger political associations such as the tonguhoe or the Chosŏn Women Comrades Society. By 1926, the Chungang yŏja ch'ŏngnyŏn yŏnmaeng (Central Alliance for Young Women) was established to unite leftist women's groups under collective leadership, but with increasing surveillance by Japanese authorities, the association dissolved soon after 1927. Park Yongock, "The Women's Modernization Movement in Korea," in Sandra Mattielli, ed., *Virtues in Conflict: Tradition and the Korean Woman Today* (Seoul: Royal Asiatic Society, Korea Branch, Samhwa Publishing Co., 1977), 107–8.

28. In view of the severity of work environments in 1930s Korea, it is important to recognize that these appeals were not demanding in the least. Their larger aims of eight-hour workdays and the abolition of child labor were policies already installed in Japan by the 1923 enactment of the Factory Law of 1911. Their visions of such benefits as maternity leave and paid holiday clearly show that colonial workers and activists engendered the designs later employed in the more consequential labor movement of the 1970s. Yi Kit'aek, *Chŏnnam tongmaeng kangnyŏng* [Platform of the Chŏllanamdo League], *Shisō geppō* 4:1 (April 1934) 36–39; Dae-sook Suh, *Documents of Korean Communism, 1918–1948*. Princeton, NJ: Princeton University Press, 1970, 171–76.

29. As Shin describes, the more "radical and violent protests led by red peasant unions, on the other hand, were centered in the northeastern regions with a high ratio of owner-cultivators, who demonstrated against local government tax policy and police interference in village affairs." Gi-Wook Shin, *Peasant Protest and Social Change in Colonial Korea* (Seattle: University of Washington Press, 1996), 6.

30. Present-day Seoul, or Hansŏng during the Chosŏn Dynasty (1392–1910), was called Kyŏngsŏng (Keijō) during the Japanese colonial period (1910–45). Dae-sook Suh, *The Korean Communist Movement, 1918–1948* (Princeton, NJ: Princeton University Press, 1967), 140; Suh, *Documents of Korean Communism*, 113–17.

31. Yi Hyojae, "Ilcheha ŭi yŏsŏng nodong munje" [The problems of female labor under Japanese rule] in An P'yŏngjik, Cho Kijun, eds., *Han'guk nodong munje ŭi*

*kujo* (Seoul: Kwangmingsa, 1980), 131–79; *Han'guk ŭi yŏsŏng undong: ŏje wa onŭl* [The Korean women's movement: yesterday and today] (Seoul: Chŏngwusa, 1989); *Yŏsŏng ŭi sahoe ŭisik* [Women's social consciousness] (Seoul: P'yŏngminsa, 1980); Kang Isu, "Kongjang ch'eje wa nodong kyuyul" [The factory system and labor control], in Kim Chin'gyun and Chŏng Kŭnsik, eds., *Kŭndae chuch'e wa singminji kyuyul kwŏllŏk* [Modernity and colonial disciplinary power] (Seoul: Munhwa kwa haksa, 1997), 117–69; "1930 nyŏndae myŏnbang taegiŏp yŏsŏng nodongja ŭi sangt'ae ŭi taehan yŏn'gu" [The conditions of female workers in the cotton textile industry of the 1930s], unpublished Ph.D. dissertation, Ewha Woman's University, 1992; Yi Chŏngok, "Ilcheha kongŏp nodong esŏ ŭi minjok kwa sŏng" [Gender and nationality in industrial labor under Japanese rule], unpublished Ph.D. dissertation, Seoul National University, 1990.

32. Although the term "light industry" has been used with reference to cotton textiles and silks, and "heavy industry" with technologically advanced enterprises, industries not only varied according to the commodity produced but differed in scale. Definitions remain imprecise, but light industries in colonial Korea were the earliest modern businesses centered on the production of consumer goods such as cotton textiles, silk, leather, paper, ceramics, lumber, printing, and the processing of rice, food, and tobacco, as well as the projects named "miscellaneous" (*kit'a*). Heavy industries focused on the manufacture of capital goods such as machinery, metals, gas, and electrical products. William N. Parker, "Europe in an American Mirror: Reflections on Industrialization and Ideology," in Richard Sylla and Gianni Toniolo, eds., *Patterns of European Industrialization: The Nineteenth Century* (London: Routledge, 1991), 84; Andrew Gordon, *The Evolution of Labor Relations in Japan; Heavy Industry, 1853–1955* (Cambridge, MA: Harvard University Press, 1988), 8–9; Sang-Chul Suh, *Growth and Structural Changes in the Korean Economy, 1910–1940* (Cambridge, MA: Harvard University Press, 1978), 106.

33. Yi Hyojae, "Ilcheha Han'guk yŏsŏng nodong munje yŏn'gu" [The problems of female labor under Japanese rule] in Yun P'yŏngsŏk, Sin Yongha, An P'yŏngjik, eds., *Han'guk kŭndae saron*, vol. 3 (Seoul: Chisik sanŏpsa, 1977), 142.

34. Over 180 labor associations outside of communist influence were formed in the mere one-year span between 1925 and 1926. An extensive study of newspaper reports of strikes from the 1920s to the 1930s indicates that labor unions and communist organizations primarily influenced general strikes. Labor demonstrations with fewer than 500 participants more frequently arose without union organization. Kim Yunhwan, *Han'guk nodong undongsa, Ilcheha p'yŏn* [History of the Korean labor movement under Japanese rule] (Seoul: Ilchogak, 1970), 177–83.

35. Thompson does not call traditional forms of social distinction *class* culture but, in detailing gentry-pleb relations and plebian culture, shows that the incipient seeds for class formation were visible even in pre-Chartist eighteenth-century English society. E. P. Thompson, *The Making of the English Working Class* (New York: Pantheon,

1963), 9; "Eighteenth-Century English Society: Class Struggle Without Class?" *Social History* 3:2 (May 1978), 163–64.

36. E. P. Thompson, "The Peculiarities of the English," in *The Poverty of Theory and Other Essays* (New York: Monthly Review Press, 1978), 296.

37. Ira Katznelson, *Working Class Formation: Nineteenth-Century Patterns in Western Europe and the United States* (Princeton, NJ: Princeton University Press, 1986), 14–23.

38. Throughout this work, I conceptualize women's work-related identities as labor consciousness, or consciousness of their skill, which refers to the methods of identification people held to specific and changing tasks rather than class alliances. Although "class" is mentioned in colonial Korean literature, the word more often signified the upper and middle classes, rather than a coherent working class. Pyŏl Moe (pseud.), "Nojajŏn ŭi nalyoe ŭi imhan chungsan kyekŭp ŭi changnae" [Concerning the divide between labor and capital: The future of the middle class], *Kaebyŏk* 18:2 (February 1926), 19–26.

39. As exemplified by Park's study of male workers in a heavy industry, in July 1938, the "Federation of the National Spiritual Mobilization Onoda Sŭnghori Branch" was organized to "promote the spiritual education of workers by emphasizing austerity and sacrifice during wartime." Soon-Won Park, "The Emergence of a Factory Labor Force in Colonial Korea: A Case Study of the Onoda Cement Factory," in Shin and Robinson, *Colonial Modernity*, 225; *Colonial Industrialization*, 151–60.

40. More intimate analysis unveils the multiple alliances of colonial Koreans. An instrumental member of a communist group in Seoul, Hŏ Maria, for example, was baptized into the Presbyterian Church before joining the Communist Party. Kim Kyŏngil, *Yi Chaeyu yon'gu: 1930 nyŏndae Seoul ŭi hyŏngmyŏngjŏk nodong undong* [A study of Yi Chaeyu: The revolutionary labor movement of 1930 Seoul] (Seoul: Ch'angjakkwa pip'yŏngsa, 1993) 63–64.

41. Han Kiyŏng, who worked at Japan Spinning and Weaving (Tae Ilbon pangjik chusik hoesa; Dai Nippon bōseki kabushiki kaisha), proudly recalled that she was the fastest spinner in her section. Han is among the nine informants whose taped interviews, from 40 to 120 minutes, are mentioned in this work. Han Kiyŏng, June 11, 1999; Kang Pokchŏm, June 3, 1999; Kim Chŏngnam, June 5, 1999; Kim Chŏngmin, June 8, 1999; Kim Ŭnnye, June 12, 1999; Kim Yŏngsŏn [male clerk], Interview conducted by Kang Isu, June 27, 1991; Pak Sun'gŭm, June 8, 1999; Yi Chaeyun, Interview conducted by Kang Isu, October 4, 1991; Yi Chungnye, June 3, 1999. See also Slavoj Žižek, "Class Struggle or Post-Modernism? Yes, Please!" in Judith Butler, Ernesto Laclau, and Slavoj Žižek, eds., *Contingency, Hegemony and Universality: Contemporary Dialogues on the Left* (London: Verso, 2000), 90–135.

42. As suffragists in the United States and England termed their cause the "woman's movement" or "woman movement" to symbolize the unity of women, the

naming of Ewha *Woman's* University in 1886 shows the adoption of such Western ideals. Elite Western suffragists built the first women's educational institution, Ewha haktang, with the support of the Chosŏn court. Kim, *Women of Korea*, 260. Nancy F. Cott, *The Grounding of Modern Feminism* (New Haven, CT: Yale University Press, 1987), 3.

43. Gerda Lerner, *The Creation of Feminist Consciousness: From the Middle Ages to Eighteen-Seventy* (New York: Oxford University Press, 1993), 14; *The Creation of Patriarchy* (New York: Oxford University Press, 1986), 198.

44. Kate Millett, *Sexual Politics* (1970; reprint, New York: Simon & Schuster, 1990); Shulamith Firestone, *The Dialectic of Sex: The Case for Feminist Revolution* (New York: Morrow, 1970).

45. Friedrich Engels, *The Origin of the Family, Private Property and the State* (New York: Pathfinder, 1972); Juliet Mitchell, *Woman's Estate* (New York: Penguin, 1971); Leopoldina Fortunati, *The Arcane of Reproduction: Housework, Prostitution, Labor and Capital*, Hilary Creek, trans. (New York: Autonomedia, 1995); Claude Meillassoux, *Maidens, Meal and Money: Capitalism and the Domestic Community* (New York: Cambridge University Press, 1981).

46. Judith Butler, *Gender Trouble: Feminism and the Subversion of Identity* (New York: Routledge, 1999), 3–7; "Contingent Foundations: Feminism and the Question of Post-Modernism," and Christina Crosby, "Dealing with Differences," in Judith Butler and Joan W. Scott, eds., *Feminists Theorize the Political* (London: Routledge, 1992), 3–21, 130–43.

47. To gradually alter the preexisting roles assigned to men and women, protofeminists of varying regions worked in progressive stages: first, to achieve social influence through the extension of domesticity; second, to gain political suffrage; third, to attain economic equality and, finally, to alter cultural definitions of womanhood. Joan W. Scott, *Gender and the Politics of History* (New York: Columbia University Press, 1988) and Nancy F. Cott and Elizabeth H. Pleck, eds., *A Heritage of Her Own* (New York: Simon & Schuster, 1979), 11–13.

48. William H. Sewell, Jr., "Toward a Post-materialist Rhetoric for Labor History," in Lenard L. Berlanstein, ed., *Rethinking Labor History* (Urbana: University of Illinois Press, 1993), 30–31.

49. The well-known phrase was first expressed in the 1969 Redstockings Manifesto, which outlined the beliefs and aims of a radical faction of the 1960s feminist initiative. Following the precedents of poststructuralist feminists, I use "personal" here and throughout to refer to those networks and realms, such as families and communities, that are closest and most familiar to the invidual.

50. The family has been theorized by a variety of scholars. Peter Berger cites the family and religious communities as the central institutions that connect the individual to society, whereas Jacques Donzelot connects the state's strengthening control over the family in eighteenth- and nineteenth-century France to the rise of philanthropy,

mass education, public work, and psychiatry. Peter L. Berger, *The Capitalist Revolution: Fifty Propositions about Prosperity, Equality, Life and Liberty* (New York: Harper Collins, 1986), 113; Jacques Donzelot, *The Policing of Families*, Robert Hurley, trans. (Baltimore, MD: Johns Hopkins Press, 1979); Eli Zaretsky, *Capitalism, the Family and Personal Life* (New York: Harper & Row, 1976), 14–15.

51. Donna Haraway, "Situated Knowledges: The Science Question in Feminism and the Privilege of Partial Perspectives," *Feminist Studies* 14:3 (Fall 1988), 586. For more on subalternity, see Dipesh Chakrabarty, *Rethinking Working-Class History: Bengal, 1890–1940* (Princeton, NJ: Princeton University Press, 1989); Partha Chatterjee, *The Nation and Its Fragments: Colonial and Postcolonial Histories* (Princeton, NJ: Princeton University Press, 1993).

52. My methodological approach draws on the findings of many scholars including Gareth Stedman Jones who asserts that the consciousness of "ideology, usually formed by a minority," did not produce politics, but politics produced ideology. Gareth Stedman Jones, *Languages of Class: Studies in English Working Class History, 1832–1982* (Cambridge: Cambridge University Press, 1983), 19.

53. Gail Hershatter, *The Workers of Tianjin, 1900–1949* (Stanford, CA: Stanford University Press, 1986), 4.

54. Chatterjee, *The Nation and Its Fragments*, 237.

55. Carlo Ginzberg, *The Cheese and the Worms: The Cosmos of a Sixteenth-Century Miller*, John and Anne Tedeschi, trans. (Baltimore, MD: Johns Hopkins University Press, 1976), xiv–xvi. For more on *high* culture, see Pierre Bourdieu, *Distinction: A Social Critique of the Judgment of Taste*, Richard Nice, trans. (Cambridge, MA: Harvard University Press, 1984); Berger, *The Capitalist Revolution*, 51.

56. Linda Alcoff, "The Problem of Speaking for Others," *Cultural Critique* 20 (1991–1992), 22–26; Pak Sŏngch'ŏl, ed., *Kyŏngbang ch'ilsimnyŏnsa* [Seventy-year history of Kyŏngbang] (Seoul: Kyŏngsŏng pangjik chusik hoesa, 1989), 55.

57. *Sinmun chŏlbal* [Scrapbooks of newspaper clippings] (Seoul: Kugwansŏ tosŏgwan [Archive of former government records] Seoul National University Central Libary, Special Collections, 1926–45), Unpublished records; *Kyŏngsŏng pangjik chusik hoesa chonp'yo* [Records of transactions of the Kyŏngsŏng Spinning and Weaving Company], Yongin, Unpublished materials, Kyŏngsŏng pangjik chusik hoesa ch'amgo, 1940–1944; *Kyŏngsŏng pangjik chusik hoesa chuju ch'onghoerok* [Records of the resolutions of the Board of Directors of the Kyŏngsŏng Spinning and Weaving Company], Yongin, Unpublished materials, Kyŏngsŏng pangjik chusik hoesa ch'amgo, 1939–1947; *Kyŏngsŏng pangjik chusik hoesa kyŏlŭirok* [Records of the resolutions of the Board of Directors of the Kyŏngsŏng Spinning and Weaving Company], Yongin, Unpublished materials, Kyŏngsŏng pangjik chusik hoesa ch'amgo, 1938–1951.

58. In spite of the sheer extent of Japanese statistical records, as contemporary Andrew Grajdanzev stated, overestimation was extremely problematic in the early years of imperial rule, tapering but continuing throughout the colonial era. Additionally, the

figures in statistical yearbooks and census records differed because the former ex-cluded analysis of household industries. Andrew J. Grajdanzev, *Modern Korea* (1944; reprint, New York: Octagon Books, 1978), 72–75.

59. The first of the nine informants serendipitously arrived at the headquarters of the Society for the Survivors and the Bereaved Families of the Pacific War (T'aep'y-ŏngyang chŏnjaeng hŭisaengja yujokhoe) after a Buddhist pilgrimage. Dressed in a monk's garb, she came to Seoul "following her instincts," unaware of my letter waiting for her at her home in Pusan. T'aep'yŏngyang chŏnjaeng hŭisaengja yujokhoe, "Kŭllo chŏngsindae hyŏnhwang" [Present conditions of the Women's Volunteer Corps], Un-published records, Seoul, Society for the Bereaved Families of the Pacific War.

60. Han Kiyŏng, June 11, 1999; Kang Pokjŏm, June 3, 1999; Kim Chŏngnam, June 5, 1999; Kim Chŏngmin, June 8, 1999; Kim Ŭnnye, June 12, 1999; Kim Yŏng-sŏn, Interview conducted by Kang Isu, June 27, 1991; Pak Sun'gŭm, June 8, 1999; Yi Chaeyun, Interview conducted by Kang Isu, October 4, 1991; Yi Chungnye, June 3, 1999.

61. Both persons were deceased by 1999.

62. Kim Chŏngmin worked at the Fujikoshi Steel Factory in Toyama, Japan. Kim Chŏngmin, June 8, 1999.

63. The demonstrators appealed reduction of working hours and pay raises. The Rubber Workers' General Strike in Seoul, which erupted in June of 1923, involved Haedong, Noegu, Tongyang, Hansŏng, and Kyŏngsŏng rubber factories (*komu kong-jang*). Although rubber production had frequently been defined as a heavy industry, implying that its labor force was male, newspaper reports chronicled the "cries of sor-row of female workers in rubber factories" during the strikes. *Chosŏn ilbo*, June 21, 1923, June 28, 1923, July 3, 4, 5, 1923; *Tonga ilbo*, July 3, 4, 1923.

NOTES TO CHAPTER I

1. Pak Hojin, "Yŏjikkong pangmun'gi" [Survey of women workers], *Kŭnu* no. 1 (May 1929), 72.

2. Wanda F. Neff, *Victorian Working Women* (New York: Columbia University Press, 1929), 42, as cited in Joan W. Scott and Louise A. Tilly, *Women, Work and Family* (New York: Routledge, 1978), 63–64.

3. Charles Benoist, *Les Ouvrières de l'aiguille à Paris: Notes pour l'étude de la ques-tion sociale*, 1895, as cited in Scott and Tilly, *Women, Work and Family*, 116.

4. My use of "development" here and throughout refers to the ways in which economies transition from the agricultural to the industrial mode of production. Karl Marx, *Capital*, vol. 1 (London: Penguin, 1976), 91.

5. The figures cited are compiled from the Government-General Statistical Yearbooks (Chōsen sōtokufu tōkei nenpō), which did not take into account house-hold industries (*kanae sugongŏp*) that hired five or fewer employees. Chōsen sōtokufu, *Chōsen sōtokufu tōkei nenpō* [Statistical yearbook of the Government-General of

Korea] (Keijō: Chōsen sōtokufu, 1922, 1933). For a table of figures on female factory workers from 1910–36, see Yi Hyojae, "Ilcheha Han'guk yŏsŏng nodong" [The problems of female labor under Japanese rule] in Yun P'yŏngsŏk, Sin Yongha, An P'yŏngjik, eds., *Han'guk kŭndae saron*, vol. 3 (Seoul: Chisik sanŏpsa, 1977), 105.

6. Post-Liberation scholars have also associated women's wage work with textile labor. Pak Chaeŭl, "Ilcheha myŏnbang chigŏp ŭi sajŏk chŏn'gae" [The historical development of the cotton textile industry under Japanese rule], unpublished Ph.D. dissertation, Kyŏnghee University, 1980.

7. Kathleen Canning, *Languages of Labor and Gender: Female Factory Work in Germany, 1850–1914* (Ithaca, NY: Cornell University Press, 1996), 283–84.

8. Scott and Tilly, *Women, Work and Family*, 63–68.

9. E. Patricia Tsurumi, *Factory Girls: Women in the Thread Mills of Meiji Japan* (Princeton, NJ: Princeton University Press, 1990), 67; Scott and Tilly, *Women, Work and Family*, 63–88; Chu Ikchong, "Ilcheha Pyongyang ŭi meriasŭ kongŏp e kwanhan yŏn'gu" [A study of the knitting industry in colonial Pyongyang], Unpublished Ph.D. dissertation, Seoul National University, 1994, 185–86.

10. Although many women workers of New England were recruited and housed, other forms of labor mobilization existed. As Thomas Dublin notes, whereas Rhode Island mills recruited "entire families," the "Waltham-Lowell firms relied upon a workforce composed chiefly of young, single women from the neighboring countryside." Similarly, Tamara Hareven concludes that whereas the first generation of workers of Manchester, New Hampshire, was composed of groups of families, the second and third generations were comprised of immigrant women. Thomas Dublin, *Farm to Factory: Women's Letters, 1830–1860* (New York: Columbia University Press, 1981), 6; *Women at Work: The Transformation of Work and Community in Lowell, Massachusetts, 1826–1860* (New York: Columbia University Press, 1981); Tamara Hareven, *Family Time and Industrial Time: The Relationship between the Family and Work in a New England Industrial Community* (Cambridge: Cambridge University Press, 1978).

11. "Boardinghouse regulations," Middlesex Company, Lowell, Massachusetts, 1846, as cited in Dublin, *Farm to Factory*, 11.

12. Lois Larcom, a widow, found refuge in the boardinghouses of the Lawrence Corporation, whereas Mary Paul lived in company housing during her first months as an apprentice. Dublin, *Farm to Factory*, 25.

13. In Korea as in Japan, "restriction to leave the premises or visit home, as well as writing and receiving letters was strictly prohibited." U Sunok, "Yŏgong ilgi" [Diary of a factory girl], *Pyŏlgŏn'gon* (March 1930), 72–73.

14. Emily Honig, *Sisters and Strangers: Women in the Shanghai Cotton Mills, 1919–1949* (Stanford, CA: Stanford University Press, 1986), 94–119.

15. Gail Hershatter, *The Workers of Tianjin, 1900–1949* (Stanford, CA: Stanford University Press, 1986), 166.

16. Meiji industrialization refers to Japan's economic development in the 1880s and 1890s, at the height of the Meiji emperor's reign (1868–1912). Tsurumi, *Factory Girls*, 142; Yasue Aoki Kidd, *Women Workers in the Japanese Cotton Mills, 1880–1920* (Ithaca, NY: Cornell University China-Japan Program, 1978), xi.

17. Alexander Gerschenkron defined many of the characteristics of *backwardness* but was, ultimately, reluctant to endorse universal models. However significant, the timing of a country's investment in modern manufacturing was not the only factor involved in shaping its industrialization. Succeeding scholars have expanded Gerschenkron's theory of developmental relativity, and recent analyses of European cases indicate that the Industrial Revolution itself in England was less dramatic than previously conceived. Alexander Gerschenkron, *Economic Backwardness in Historical Perspective: A Book of Essays* (Cambridge, MA: Belknap Press, 1966), 33, 355–56. N. F. R. Crafts, S. J. Leybourne, and T. C. Mills, "Britain," in Richard Sylla and Gianni Toniolo, eds. *Patterns of European Industrialization: The Nineteenth Century* (London: Routledge, 1991), 109–52.

18. Robert Wade, *Governing the Market: Economic Theory and the Role of Government in East Asian Industrialization* (Princeton, NJ: Princeton University Press, 1990), 351.

19. Andre Schmid, *Korea Between Empires: 1895–1919* (New York: Columbia University Press, 2002), 10.

20. The Chosŏn government officially requested Chinese aid, whereas Japanese militias were unwarranted.

21. Alexis Dudden, *Japan's Colonization of Korea: Discourse and Power* (Honolulu: University of Hawaii Press, 2005), 2; Hilary Conroy, *The Japanese Seizure of Korea, 1868–1910: A Study of Realism and Idealism in International Relations* (Philadelphia: University of Pennsylvania Press, 1960).

22. Catherine Hall and Sonia Rose, eds., *At Home with the Empire: Metropolitan Culture and the Imperial World* (Cambridge: Cambridge University Press, 2006), 7.

23. Edward I-te Chen, "Japanese Colonialism: An Overview," in Harry Wray and Hilary Conroy, eds., *Japan Examined: Perspectives on Modern Japanese History* (Honolulu: University of Hawaii Press, 1983), 204.

24. Andrew Grajdanzev, *Modern Korea* (1944; reprint, New York: Octagon Books, 1978), 238–49; Dudden, *Japan's Colonization of Korea*, 131–37. See also Patrick Joyce, *The Rule of Freedom: Liberalism and the Modern City* (London: Verso, 2003), 3.

25. Michel Foucault, "Governmentality," in Graham Burchell, Colin Gordon, and Peter Miller, eds., *The Foucault Effect* (Chicago: University of Chicago Press, 1991), 87–104.

26. Nikolas Rose, *Governing the Soul: The Shaping of the Private Self* (London: Free Association, 1989), xx.

27. Anthony Giddens, *The Constitution of Society: Outline of the Theory of Structuration* (Berkeley: University of California Press, 1984), xxxiii, 162.

28. Schmid, *Korea Between Empires*, 13; Hŏ Changman, "1920 nyŏndae minjok kaeryang chuŭi ŭi kaekŭpchŏk kich'o haemyŏng esŏtoenŭn myŏt kaji munje"

[Fundamental problems of the progressive nationalism of the 1920s], *Ryŏksa kwahak* [Historical Science] (March 1966), 37–43; Kim Hŭiil, "Minjok kaeryang chuŭi ŭi keykŭpchŏk kich'o nŭn yesok purŭjyoji yida" [The class origins of nationalist progressivism is bourgeoise], *Ryŏksa kwahak* (April 1966), 38–46; Kim Indŏk, *Singminji sidae chaeil Chosŏnin undong yŏn'gu* [A study of the movements of Koreans in Japan during the colonial era] (Seoul: Kukhak charyowŏn, 1996); Michael E. Robinson, *Cultural Nationalism in Colonial Korea, 1920–1925* (Seattle: University of Washington Press, 1988).

29. Peter Duus, *The Abacus and the Sword: The Japanese Penetration of Korea, 1895–1910* (Berkeley: University of California Press, 1995); "Economic Dimensions of Meiji Imperialism, 1895–1910," in Ramon H. Myers and Mark R. Peattie, *The Japanese Colonial Empire, 1895–1945* (Princeton, NJ: Princeton University Press, 1984), 128–71.

30. Ki-baek Lee (Yi Ki-baek) and Edward Wagner, trans., *A New History of Korea* (Cambridge, MA: Harvard University Press, 1984), 281–315; Sang-chul Suh, *Growth and Structural Changes in the Korean Economy, 1910–1940* (Cambridge, MA: Harvard University Press, 1978), 101.

31. Walt Rostow termed the various points in the shift as "stages of development" that were universally apparent. Walter W. Rostow, *The Stages of Economic Growth: A Non-Communist Manifesto* (1960; reprint, New York: Cambridge University Press, 1990), 161–62.

32. Joyce, *The Rule of Freedom*, 233.

33. Chulwoo Lee, "Modernity, Legality and Power in Korea under Japanese Rule," in Gi-Wook Shin and Michael Robinson, eds., *Colonial Modernity in Korea* (Cambridge, MA: Harvard University Press, 1999), 26; Gi-Wook Shin, *Peasant Protest and Social Change in Colonial Korea* (Seattle: University of Washington Press, 1996), 40–42.

34. Nevertheless, it is important to remember that manufacturing enterprises existed in Korean cities before significant colonial investment. Chōsen sōtokufu, *Keijō sangkōgyō chōsa* [Industries in Kyŏngsŏng] (Keijō: Chōsen sōtokufu, 1913); Martina Deuchler, *Confucian Gentlemen and Barbarian Envoys: The Opening of Korea, 1875–1885* (Seattle: University of Washington Press for the Royal Asiatic Society, Korea Branch, 1977), 73–74. See also Suh, *Growth and Structural Changes*, 5; Carter J. Eckert, *Offspring of Empire: The Koch'ang Kim's and the Colonial Origins of Korean Capitalism, 1876–1945* (Seattle: University of Washington Press, 1991), 270n21, 41.

35. Since its enactment, the Company Law suffered from much criticism for its uncertain grounds on the integration of Japanese and Korean law. It was revised twice between 1911 and the point of its abolition in 1920. Manufacturing burgeoned thereafter and as Carter Eckert describes, the "existing tariff barriers between Korea and Japan were also largely eliminated after the war, permitting the free flow into Korea of the Japanese capital goods so essential to the establishment

of modern machine factories." Eckert, *Offspring of Empire*, 38–43; Suh, *Growth and Structural Changes*, 7; Grajdanzev, *Modern Korea*, 92.

36. By 1920, Japan paid off more than 1 billion yen in debt and established a credit nearing 2.8 billion yen. Nakamura Takafusa, *Economic Growth in Prewar Japan*, Robert A. Feldman, trans. (New Haven, CT: Yale University Press, 1983), 152, as cited in Carter J. Eckert, "Total War, Industrialization and Social Change in Late Colonial Korea," in Peter Duus, Ramon H. Myers, and Mark R. Peattie, eds., *The Japanese Wartime Empire, 1931–1945* (Princeton, NJ: Princeton University Press, 1996), 6, 40–69.

37. Sŏng T'aegyŏng, "Samil undong sigi ŭi Han'guk nodongja ŭi hwaltong e tae-hayŏ" [Concerning the activities of Korean workers in the March First era], *Yŏksa hakpo*, no. 41 (March 1969), 52–83.

38. One million yen in the late 1910s was equivalent to 400,000 US dollars (USD) in 1920, the contemporary value of which is around 4.6 million USD. Chōsen sōtokufu, *Chōsen sōtokufu tōkei nenpō* [Statistical yearbook of the Government-General of Korea] (Keijō: Chōsen sōtokufu, 1911, 1921, 1930); Soon-Won Park, *Colonial Industrialization and Labor in Korea: The Onoda Cement Factory* (Cambridge, MA: Harvard University Press, 1999), 51–114; Suh, *Growth and Structural Changes*, 11; Eckert, *Offspring of Empire*, 42.

39. In 1926 Noguchi Jun financed both companies to tap into the vast hydro-electric resources available on the peninsula. An P'yŏngjik, "Nihon chissū ni okeru Chōsenjin rōdōsha kaikyū no seicho ni kansuru kenkyū" [Korean Workers and the Japanese Nitrogen Fertilizer Company], *Chōsenshi kenkyūkai ronbunshū*, 25 (1988), 157–92; Chŏn Sŏktam, Kim Hanju, and Yi Kisu, *Ilcheha ŭi Chosŏn sahoe kyŏngjesa* [The social and economic history of Korea under Japanese rule] (Seoul: Hyŏptong mun'go chogŭm yŏnp'an, 1947), 101–3; Kim Yunhwan, *Han'guk nodong undongsa, Ilcheha pyŏn* [History of the Korean labor movement under Japanese rule] (Seoul: Ilchogak, 1970), 53–54.

40. Suh, *Growth and Structural Changes*, 1–13; Park, *Colonial Industrialization*, 21–33; An, "Nihon chissū," 499–549.

41. Especially concerning manufacturing workers, vast discrepancies prevail between the two sources. In statistical yearbooks, surveying only businesses with five employees or more accounted for approximately 85,000 factory workers in 1930. But in the 1930 census that included household industries, the total number of manufacturing workers was closer to 285,000. The Government-General performed five surveys, in 1925, 1930, 1935, 1940, and 1944. *Chōsen kokusei chōsa hōkoku* [National census of the Government-General of Korea] (Keijō: Chōsen sōtokufu); Paek Ugin, "Singminji sidae kyekŏp kujo ŭi kwanhan yŏn'gu" [A study of the class structure during the colonial era], in Han'guk sahoesa yŏn'guhoe, ed., *Singminji siodae kyekŭp kujo ŭi kwanhan yŏn'gu* [Social class and social change in Korean society], Han'guk sahoesa yŏn'guhoe nonmunjip, vol. 8 (Seoul: Munhak kwa chisŏngsa, 1987), 121–245.

42. The rates of population growth in the early years of colonial rule were as high as 3.05 to 5.58 percent a year, indicating that the figures of the early statistical yearbooks underestimated the number of Koreans by at least two million. According to economist Simon Kuznets' estimates, the growth rates of populations during early industrialization rarely exceeded 1.5 percent per annum. Though Grajdanzev showed that Korea's rate was not as high as some contemporary Latin American countries, it exceeded Japan's population growth that was estimated at about 1 percent per annum from 1869 to 1910. The Japanese population in Korea likewise increased from 172,000 (1.3 percent) in 1910 to 650,000 (2.9 percent) in 1939. Grajdanzev, *Modern Korea*, 72–73; Alice H. Amsden, *Asia's Next Giant: South Korea and Late Industrialization* (Oxford: Oxford University Press, 1989), 190–91.

43. Chōsen sōtokufu, *Chōsen kokusei chōsa hōkoku* (Keijō: Chōsen sōtokufu, 1930, 1940); Suh, *Growth and Structural Changes*, 50–51. Shin, *Peasant Protest*, 44.

44. Kim Yŏngmo, "Kŭnaehwa ŭi sagak: nodong munje" [The dead angle of modernization: The problem of labor], *Sedae* (November 1971), 169–243; Kim Yŏngsŏp, "Hanmal-Ilcheha ŭi chijuje: Kanghwa Kimssiga ŭi ch'usugi rŭl t'onghaesŏ pon chiju kyŏngyŏng" [The landlord system of the late Chosŏn dynasty and under colonial rule: A case study of the Kanghwa Kim family], *Tonga munhwa* 11 (November 1972), 3–86.

45. Gi-Wook Shin notes that the number of landlord agents grew from 18,785 to 33,195 between the mid-1920s and 1930. Small-scale landlord-cultivators were "three to five times more common than" larger, commercial landlords. Shin, *Peasant Protest*, 46–47. For details of household production in a rural economy, see Clark W. Sorensen, *Over the Mountains Are Mountains: Korean Peasant Households and Their Adaptations to Rapid Industrialization* (Seattle: University of Washington Press, 1988), 127–58.

46. Soon-Won Park explains that Seoul's population boom was caused in part by the expansion of the administrative boundaries of the capital in April 1936. The territory "increased by a factor of 3.8 and the population by a factor of 1.7." The growth of Seoul also contributed to the growth of female labor-intensive enterprises and the number of women workers. By 1928, the number of cotton textile and rubber factories hiring women rose to thirty-three. Park, *Colonial Industrialization*, 17; Grajdanzev, *Modern Korea*, 80; Hori Kazuō, "1930 nyŏndae sahoejŏk punŏp ŭi chaep'yŏnsŏng: Kyŏnggido: Kyŏngsŏngbu ŭi punsŏk ŭl t'onghayŏ" [The reorganization of the social division of labor in the 1930s: An analysis of Kyŏnggi Province and Kyŏngsŏng] in An P'yŏngjik and Nakamura Tetsu, eds., *Kŭndae Chosŏn kongŏphwa ŭi yŏn'gu* [A study of industrialization in modern Korea] (Seoul: Ilchogak, 1993), 53; Chōsen sōtokufu, *Chōsen kokusei chōsa hōkoku* [National census of the Government-General of Korea] (Keijō: Chōsen sōtokufu, 1940); "Kyŏngsŏng yŏjikkong t'onggye" [Statistics of women workers in Kyŏngsŏng], *Chosŏn chigwang*, 78 (May 1928), 51–52.

47. In Suh's study, "southern provinces" signified Kyŏnggi, North and South Ch'ungch'ŏng, North and South Chŏlla, and North and South Kyŏngsang, whereas the northern provinces were Hwanghae, North and South Hamgyŏng, and North and South P'yŏngan. Suh, *Growth and Structural Changes*, 134–35; Chŏngch'ijŏk kyŏlsa wa undong e taehayŏ [On political associations and movements], *Tonga ilbo*, January 29, 1924; Keijō shōkō kaigisho [Kyŏngsŏng chamber of commerce and industry], *Keijō shōko kaigisho tokei nenpō* [The annual statistical report of the Kyŏngsŏng chamber of commerce and industry] (Keijō: Keijō shōkō kaigisho, 1930, 1935).

48. An Pyŏngjik and Hori Kazuō, "Singminji kongŏphwa ŭi yŏksajŏk chogŏn kwa kŭ sŏnggyŏk" [The historical conditions and character of colonial industrialization in Korea], in An and Nakamura, eds., *Kŭndae Chosŏn kongŏphwa*, 37; Park, *Colonial Industrialization*, 19.

49. Chŏng Hyegyŏng, "Ilcheha chae-Il Han'gukin minjok undong ŭi yŏn'gu: Taep'an chibang ŭl chungsin ŭro" [A study of the nationalist movements of Koreans in Japan under Japanese rule: A focus on the Ōsaka region], unpublished Ph.D. dissertation, Han'guk chŏngsin munhwa yŏn'guwŏn (Academy of Korean Studies), 1998, 19–32; Kim Minyŏng, *Ilche ŭi Chosŏnin nodongnyŏk sut'al yŏn'gu* [A study of the exploitation of the Korean labor movement after Liberation] (Seoul: Ilchogak, 1970); Wayne Patterson, *The Korean Frontier in America* (Honolulu: University of Hawaii Press, 1988), 1–11; Park, *Colonial Industrialization*, 19.

50. Hershatter, *The Workers of Tianjin*, 3.

51. At the Onoda Cement Factory, three categories of laborers—novices (*teigō*), apprentices (*minarai*), and regular workers (*honshokkō*)—prevailed. Park, *Colonial Industrialization*, 66–69.

52. Hershatter, *The Workers of Tianjin*, 42.

53. The Korea Waterpower Corporation was founded in 1926 and the Korea Nitrogen Fertilizer Company in 1927, but production did not start until the early 1930s. Park, *Colonial Industrialization*, 87; Suh, *Growth and Structural Changes*, 176–77. For more on the Depression and Chinese workers, see Honig, *Sisters and Strangers*, 32.

54. The Major Industries Control Law stabilized production, government controls of sales, and prices in Japan, and thus greatly facilitated Japanese investment in Korea and Taiwan. Apart from Korea's strategic significance, Japanese businessmen of the 1930s benefited from the cost-effectiveness of natural resources, such as hydroelectric power, and of employing colonial workers. Eckert, "Total War," 4.

55. Chōsen sōtokufu, *Chōsen sōtokufu tōkei nenpō* [Statistical yearbook of the Government-General of Korea] (Keijō: Chōsen sōtokufu, 1937, 1938).

56. The colony's first governor-general, Terauchi Masatake, installed the public education system. Although private schools constituted a fraction of the total number of schools in 1919, it is important to recall that these institutions provided the greatest amount of relief to students and activists during the March First demonstrations.

Notwithstanding that the educational system improved during the second half of the colonial era, inasmuch as it provided more people with the opportunity to study, it also furthered Japan's "cultural hegemony" over Korea. E. Patricia Tsurumi, "Colonial Education in Korea and Taiwan," in Ramon H. Myers and Mark R. Peattie, *The Japanese Colonial Empire, 1895–1945* (Princeton, NJ: Princeton University Press, 1984), 296; Yung-chung Kim (Kim Yungch'ŏng), ed., trans., *Women of Korea: A History from Ancient Times to 1945* (Seoul: Ewha Woman's University Press, Kwang Myŏng Printing Co., 1976), 259–61; Kim Hwallan (Helen), *Rural Education for the Regeneration of Korea* (Philadelphia: Dunlap Printing Co., 1931), 48–51.

57. During the first half of the colonial era, women who went to schools were usually privileged. This was not a consequence of the elitism of the schools but, rather, economic circumstance. In spite of the fact that Protestants boasted of equal universal education, those they educated were frequently members of the upper classes. Tsurumi, "Colonial Education," 300.

58. The number of women enrolled in colonial schools in 1918 was 12,300. The number of boys enrolled in the same year was more than 800,000. Lee Kyu-hwan (Yi Kyuhwan), "A Study of Women's Education under Japanese Occupation," *Journal of the Korean Cultural Research Institute of Ewha University* 3 (1962), 145–46; Kim Hwallan (Helen), "Chigŏp chŏnsŏn kwa Chosŏn yŏsŏng" [Professional progress and the Chosŏn woman], *Sintonga* 11 (September 1932), 2–9; Tsurumi, "Colonial Education," 305.

59. Edward P. Thompson, *The Making of the English Working Class* (New York: Pantheon, 1963), 191–92.

60. Park, *Colonial Industrialization*, 28.

NOTES TO CHAPTER 2

1. Ada Heather Biggs, as cited in Joan W. Scott and Louise A. Tilly, eds., *Women, Work and Family* (New York: Routledge, 1978), vi. The changing characteristics of women's work were similarly debated by contemporary Koreans. For more, see Pak Aa, "Yŏsŏng konghwang sidae" [Age of alarm for women], *Pyŏlgŏn'gon* (July 1932); Yi Sŏnghwan, "Puin munje kwanhwa: Puin kwa chigŏp chŏnsŏn" [Problems and solutions to the woman question: Women and the battle of work], *Sinyŏsŏng* 10:3 (March 1932), 13–19; Yi Taeŭi, "Yŏjaŭi kyŏngjejŏk tongnip" [Women's economic liberation], *Ch'ŏngnyŏn* (January 1922).

2. By *wage work* I refer to a broad range of paid work from daily and production-based wage labor to professional employment.

3. I mention "household" here and throughout to signify the familial unit, which was frequently but not always comprised of immediate kin. Friedrich Engels, *The Origin of the Family, Private Property and the State* (New York: Pathfinder, 1972); Gary S. Becker, "The Economic Way of Looking at Behavior: The Nobel Lecture," *Essays in Public Policy* 69 (Stanford, CA: Hoover Institute, 1996), 21; *A Treatise on the*

*Family* (1981; reprint, Cambridge, MA: Harvard University Press, 1991); Scott and Tilly, *Women, Work and Family*, 2.

4. Clark W. Sorensen, *Over the Mountains Are Mountains: Korean Peasant Households and Their Adaptations to Rapid Industrialization* (Seattle: University of Washington Press, 1988), 38–40; Richard R. Wilk, ed., *The Household Economy* (New York: Westview Press, 1989); Tamara Hareven, *Family Time and Industrial Time: The Relationship between the Family and Work in a New England Industrial Community* (Cambridge: Cambridge University Press, 1978).

5. Chŏng Chinsŏng, "Singminji chabon chuŭihwa kwajŏng esŏ ŭi yŏsŏng nodong ŭi p'yŏnmo" [The transfiguration of female labor during the capitalization process in colonial Korea], *Han'guk yŏsŏnghak* 4 (1988), 55; Gi-Wook Shin, *Peasant Protest and Social Change in Colonial Korea* (Seattle: University of Washington Press, 1996), 49–51; Kim Yunhwan, "Kŭndaejŏk imgŭm nodong ŭi hyŏngsŏng kwajŏng" [The formation of modern wage work], in An P'yŏngjik and Cho Kijun eds., *Han'guk nodong munje ŭi kujo* (Seoul: Kwangmingsa, 1980), 59–95.

6. Scott and Tilly, *Women, Work and Family*, 2, 104; Sorensen, *Over the Mountains*, 41; Yi Paeyong, "Han'guk kŭndae sahoe chŏnhwan kwa honin chedo ŭi p'yŏnhwa" [The transformation of modern Korean society and changes in the institution of marriage], *Ehwa saga yŏn'gu* 23 and 24 (1997), 39–54.

7. Chōsen sōtokufu, *Chōsen no keizai jijō* [Economic conditions in Korea] (Keijō: Chōsen sōtokufu, 1931, 1934); Yi Chonghyŏn "Ilche kangjŏmha 1920 nyŏndae Chosŏn nodong kyegŭp ŭi saenghwal sangt'ae" [The living conditions of the working class during Japanese occupation], *Ryŏksa kwahak* (February 1962), reprinted in Kim Kyŏngil, ed., *Puk-Han hakkye ŭi nonong undong yŏn'gu* [North Korean studies of labor and peasant movements] (Seoul: Ch'angjak kwa pip'yŏngsa, 1989), 86–113; Shin, *Peasant Protest*, 50; Scott and Tilly, *Women, Work and Family*, 2.

8. Sorensen, *Over the Mountains*, 35.

9. Carol B. Stack and Linda M. Burton, "Kinscripts: Reflections on Family, Generation and Culture," in Evelyn Glenn, Grace Chang, and Linda Forcey, eds., *Mothering: Ideology, Experience, and Agency* (London: Routledge, 1994), 33.

10. In 1925, approximately 33.6 percent of child workers under fifteen lived in cities, but by 1930, this figure had risen to 34.3 percent, and to 36.2 percent in 1935. See Chōsen sōtokufu, *Chōsen kokusei chōsa hōkoku* [National census of the Government-General of Korea] (Keijō: Chōsen sōtokufu, 1925, 1930, 1935); Yunshik Chang, "Population in Early Modernization: Korea," unpublished Ph.D. dissertation, Princeton University, 1967, 131.

11. Kun Wŏn (pseud.), "Puin munje ŭi chongjŏn" [Various women's questions], *Sinsaenghwal* 1:1 (March 1922); Kim Myŏngho, "Chosŏn yŏsŏng kwa chigŏp" [Korean women and professions], *Sinyŏsŏng* 3:10 (October 1925), 9–12; Chang, "Population in Early Modernization: Korea," 238, 257; Chŏng, "Singminji chabon chuŭihwa," 62–65.

12. Kwŏn T'aeŏk, *Han'guk kŭndae myŏnŏpsa yŏn'gu* [A study of textile industry in Korea] (Seoul: Ilchogak, 1989), 225–43; Sorensen, *Over the Mountains*, 133–34.

13. Chōsen sōtokufu, *Chōsen kokusei chōsa hōkoku* (1935); Shin, *Peasant Protest*, 135.

14. *Tonga ilbo*, April 6, 1925, April 16, 1933.

15. *Tonga ilbo*, January 20, 1933, July 21, 1933, January 23, 1936.

16. Chōsen sōtokufu gakumukyoku shakaika, *Kōjō oyobi kōzan ni okeru rōdō jōkyō chōsa* [Survey of labor conditions in factories and mines] (Keijō: Chōsen sōtokufu, 1933), 84, 233–43; Ch'oe Sukkyŏng, Yi Paeyŏng, Sin Yŏngsuk, and An Yŏnsŏn, "Han'guk yŏsŏngsa chŏngnip ŭl wihan yŏsŏng inmul yuhyŏng yŏn'gu: Samil undong ihu put'ŏ haebang kkaji" [A study of characteristics of notable women for the strengthening of Korean women's history: From the March First movement to liberation] *Yŏsŏnghak nonjip*, vol. 10 (Seoul: Ewha Woman's University and Han'guk yŏsŏng yŏn'gusoe, 1993), 50; Chŏng, "Singminji chabon chuŭihwa," 85.

17. *Tonga ilbo*, May 6, 1933, July 19, 1933; *Maeil sinbo*, June 26, 1934, December 3, 1938.

18. *Tonga ilbo*, January 20, 1933, July 21, 1933, August 10, 1933, March 19, 1935.

19. Chŏk Tosaeng, "Nongch'on yŏsŏng pit'ongsa ŭi taehwa" [On the sad history of rural women], *Chosŏn nongmin* 3:9 (1927), 21.

20. Chŏng, "Singminji chabon chuŭihwa," 75.

21. Andrew J. Grajdanzev, *Modern Korea* (1944; reprint, New York: Octagon Books, 1978), 151–52; Chōsen sōtokufu, *Chōsen no bussan* [Commodities of Korea] (Keijō: Chōsen sōtokufu, 1927).

22. *Tonga ilbo*, May 1, 1935; Yun Chohŭi, "Nongch'on kwa puŏp" [Secondary business and the countryside], *Sinmin* (June 1929), 93–94; Chŏng, "Singminji chabon chuŭihwa," 76.

23. *Tonga ilbo*, June 24, 1927, May 30, 1936; Kaname Akamatsu, "A Historical Pattern of Economic Growth in Developing Countries," *The Developing Economies* 1 (March–August, 1962), 3–25; Chŏng, "Singminji chabon chuŭihwa," 62.

24. In 1930, seasonal workers accounted for 48.3 percent of the total agricultural population. Most likely, these were young boys who served as "farmhands" during harvest seasons. Pak Kyŏngsik, *Ilbon cheguk chuŭi ŭi Chosŏn chibae* [The management of Korea during Japanese imperialism] (Seoul: Ch'ŏnga sinsŏ, 1986), 264.

25. Although 0.3 percent of domestic servants commuted (*tonggŭn*), the majority were "live-in" (*sangju*) employees. Chōsen sōtokufu, *Chōsen kokusei chōsa hōkoku* [National census of the Government-General of Korea] 1930: 224.

26. *Tonga ilbo*, March 13, 1928, November 28, 1937.

27. Kim Wŏnju, "Yeppŭni nŭn ŏdiro?" [Where do the attractive go?] *Sinyŏsŏng* 9:12 [New Woman] (December 1931), 105; *Tonga ilbo*, December 20, 1925, April 5, 1926, October 10, 1937. See also Chang Lin (pseud.), "Nodong puin ŭi chojik hwarŭl" [Concerning strategies for working women], *Kŭnu* (May 1929); Chŏk Ku (pseud.), "Yŏjikkong woesap'yŏn" [Women workers], *Kaebyŏk* 18:4 (April 1926), 110–13.

28. Kim Yŏngsŏn. Taped Interview conducted by Kang Isu, June 27, 1991; Kim Namch'ŏn, "Yŏsŏng ŭi chigŏp munje" [Women's professional problems], *Yŏsŏng*, 5:12 (1940), 22–27.

29. Soe Mi (pseud.), "Nodongja kyori mundap" [Discussions with workers], *Kaebyŏk* 18:3 (March 1926), 111–12. See also the writings of Song Kyewŏl, a frequent contributor to contemporary women's journals. Song Kyewŏl, "Yŏjik kongp'yŏng: kongjang sosik" [Edition on women workers: Factory news], *Sinyŏsŏng* 9:12 (December 1931), 107–10.

30. Sŏng T'aegyŏng, "Samil undong sigi ŭi Han'guk nodongja ŭi hwaltong e taehayŏ" [Concerning the activities of Korean workers in the March First era], *Yŏksa hakpo* 41 (March 1969), 59–61; An Tonghyŏk, "Kongŏp Chosŏn ŭi chŏnmo" [The comprehensive portrait of industrialization in Korea], *Sintonga* (June 1935), 33–46.

31. For variation, I use "spinning and weaving" and "textiles" synonymously in the translations of company names. Kyŏngsŏng Spinning and Weaving and Kyŏngsŏng Textiles, for example, are employed interchangeably to refer to the same company. Chŏng Kŭnsik, "Ilcheha Chongyŏn pangjŏk ŭi chamsaŏp chibae" [Labor control of the Chŏngyŏn (Kangegafuchi) silk-reeling factory under Japanese colonial rule], in *Han'guk kŭndae nongch'on sahoe wa Ilbon cheguk chuŭi* [Rural society and Japanese Imperialism in Modern Korea], Han'guk sahoesa yŏn'guhoe nonmunjip, vol. 2 (Seoul: Munhak kwa chisŏngsa, 1986), 159.

32. Chōsen sōtokufu gakumukyoku shakaika, *Kōjō oyobi kōzan*, 38–39; Kwŏn, *Han'guk kŭndae myŏnŏpsa*, 247–52; Keizō Seki, *The Cotton Industry of Japan* (Tokyo: Japan Society for the Promotion of Science, 1956), 4–13.

33. In 1919, Min Pyŏngsŏk founded the Chosŏn Silk Reeling Company and Pak Yŏngho founded the Kyŏngsŏng Spinning and Weaving Company. Both companies were instituted with an approximate one million yen in invested capital. Sŏng, "Samil undong sigi," 59; Kwŏn T'aeŏk, "Kyŏngsŭng chingnyu chusik hoesa ŭi sŏngnip kwa kyŏngyŏng" [The establishment and management of the Kyŏngsŏng Cord Company], *Han'guk saron* 6:12, (December 1979), 300; Sang-Chul Suh, *Growth and Structural Changes in the Korean Economy, 1910–1940* (Cambridge, MA: Harvard University Press, 1978), 98–100; Pak Inhwan, ed., *Kyŏngsŏng pangjik yuksimnyŏnsa* [Sixty-year history of the Kyŏngsŏng Spinning and Weaving Company] (Seoul: Kyŏngsŏng pangjik chusik hoesa, 1980).

34. Of approximately 17 million yen accumulated by textile production in 1929, factory output comprised 8.3 million yen whereas household industries yielded 8.7 million yen. Chōsen sōtokufu, *Chōsen no bussan* [Commodities of Korea] (Keijō: Chōsen sōtokufu, 1927), 496; Chŏn Sŏktam and Ch'oe Unkyu, *Chosŏn kŭndae sahoe kyŏngjesa* [The social and economic history of modern Korea] (1959; reprint, Seoul: Yisŏng kwa hyŏnsilsa, 1989), 232–33.

35. The Kanegafuchi Spinning and Weaving Company (Kanegafuchi bōseki kabushiki kaisha) was established in 1887 with one million yen in invested capital. By

1910, it was one of the six largest textile companies in Japan. Grajdanzev, *Modern Korea*, 154; Chŏng, "Ilcheha Chongyŏn pangjŏk," 159; Chōsen kōgyō kyōkai (Chōsen industrial association), *Chōsen no kōgyō to sono shigen* [Industry and resources of Chōsen] (Keijō: Chōsen kōgyō kyōkai, 1937). For more on Japanese textile giants in Korea, see Tōyō bōseki kabushiki kaisha [East Asia Spinning and Weaving Company], *Tōyōbō hyaku nenshi* [One hundred year history of Tōyōbō] (Tokyo: Tōyō bōseki kabushiki kaisha, 1986), 307–8; Kanegafuchi bōseki kabushiki kaisha, *Kanebo hyakunenshi* [One hundred-year history of the Kanegafuchi Spinning and Weaving Company] (Osaka: Kanegafuchi bōseki kabushiki kaisha, 1988).

36. The Kwangmyŏng Company had workshops in Kongp'yŏngdong and P'yŏngdong, the Samu Company in Kyŏnamdong and Sogyŏktong. Taechang Hosiery formed bases in the Kwanghwamun, Songhyŏndong, and Inŭidong areas. Chu Ikchong, "Ilcheha Pyongyang ŭi meriasŭ kongŏp e kwanhan yŏn'gu" [A study of the knitting industry in colonial Pyongyang], unpublished Ph.D. dissertation, Seoul National University, 1994, 74; Suh, *Growth and Structural Changes*, 98–100.

37. The Pyongyang knitting industry instituted budget cuts to compete with rival Chinese firms in Sinŭiju in the late 1920s and the early 1930s. Nonetheless, as demand increased after the Japanese takeover of Manchuria, Korean-owned knitting enterprises once again prospered throughout the 1930s. Some of the larger knitting companies in Pyongyang were Sechang, Samgong, Taedong, Taesŏng, Taesan, Wŏlsŏng, Ilsin, Tonggwang, Paekhwa, T'aeyang, Ch'ohŭng, and Chosŏn. Chu, "Ilcheha Pyongyang ŭi meriasŭ, 173–74.

38. Soon-Won Park, *Colonial Industrialization and Labor in Korea: The Onoda Cement Factory* (Cambridge, MA: Harvard University Press, 1999), 31; Chōsen sōtokufu, *Chōsen no sangyō* [Industries of Korea] (Keijō: Chōsen sōtokufu, 1933, 1935); *Chōsen sembaishi* [Industrial history of Korea], vols. 1–3 (Keijō: Chōsen sōtokufu, 1936).

39. Kang Isu, "1920–1960 nyŏn Han'guk yŏsŏng nodong sijang kujo ŭi sajŏk p'yŏnhwa: Koyong kwa imgŭm kyŏkch'a p'yŏnhwa rŭl chungsim ŭro" [Historical changes within the structure of the female labor market in Korea, 1920–1960: Changes in employment and wage differences], *Yŏsŏng kwa sahoe*, no. 4 (1993), 181; Chŏng, "Singminji chabon chuŭihwa, 65; Chŏng Osŏng, "Yisip ch'ilnyŏn kan kŭllo hadŭn yŏjikkong An Ahyŏng ŭi adhwa: Kŭka kongjang e ssŭlŏjŏ chukki kkaji" [Seventeen years of work, the sad tale of An Ahyŏng], *Sinyŏsŏng* 7:6 (June 1929), 123–43.

40. The project was launched by Noguchi Jun, with invested capital of twenty million and ten million yen for the respective companies. In 1926, a project to construct the Pujŏn River Dam was initiated, and by January 1930, the Korean nitrogen fertilizer plant started operations using power from the Pujŏn Water Power Station. Before 1933, the Japanese government in Korea did not permit the establishment of any gunpowder factories in Korea, but with increasing militarization in the late 1930s, such prohibitions were reversed. Grajdanzev, *Modern Korea*, 160–63; Park, *Colonial Industrialization*, 87–88; Suh, *Growth and Structural Changes*, 100.

41. In terms of production value, the leading chemical enterprises in 1935 were as follows: the production of fertilizers yielded an approximate 55.0 million yen; the production of sulfate ammonia, 28.2 million; the production of fish oil, 17.4 million; the production of fish fertilizers, 14.5 million; the production of rubber footwear, 10.5 million. Many owners of rubber companies in the Pyongyang area were also owners of other enterprises, with the most prominent being rice polishing (*ch'ungmi*) and hosiery (*yangmal*) and linen (*p'omok*) manufacturing. *Tonga ilbo*, January 1, 1928; *Chungwoe ilbo*, August 20, 1930; Kang Munsŏk, "Chosŏn minjok haebang undong ŭi yissŏ sŏŭi Chosŏn sanŏp nodong chosaso ŭi samyŏng" [A call of the Chosŏn industrial labor research institute], *Kaebyŏk* 22:1 (January 1930), 59–63.

42. Although the majority of rubber factories were Korean owned, Japanese-owned rubber factories existed as well. Some of the larger companies were Naitoku Rubbers in Pyongyang, Asahi Rubbers in Pusan, Nissei Rubbers in Taegu, Chōsen Rubbers in Inch'ŏn, the Chōsen Rubber Company in Seoul, and the Sanwa Rubber Company.

43. The growth rate of the rubber shoe industry was comparable to the growth rate of the fertilizer industry, including nitrogen fertilizer. Whereas in 1935, the production value of fertilizers was 55 million yen, it rose to 90.5 million yen in 1937—an increase of approximately 61 percent. *Sŏjo ilbo*, February 21, 1938; Grajdanzev, *Modern Korea*, 160–66; Chu, "Ilcheha Pyongyang ŭi meriasŭ, 108, 166.

44. Chōsen sōtokufu gakumukyoku shakaika, *Kōjō oyobi kōzan*, 316; Masahisa Kōji, "Chōsen ni okeru chūyō kōjō no rōdō jijō" [Labor situation at the major factories and mines in Korea], *Shokusan chōsa geppō* 50 (July 1942); Miya Kōichi, *Chōsen no rōdōsha* [The laborers of Korea] (Tokyo: Totsu shoseki, 1945), 44.

45. Edward P. Thompson, *The Making of the English Working Class* (New York: Pantheon, 1963), 416.

46. Scott and Tilly, *Women, Work and Family*, 119, 229.

47. James Scott, *The Moral Economy of the Peasant* (New Haven, CT: Yale University Press, 1976), 1–5; *Weapons of the Weak: Everyday Forms of Peasant Resistance* (New Haven, CT: Yale University Press, 1985); Samuel Popkin, *The Rational Peasant* (Berkeley: University of California Press, 1979), 17–21.

NOTES TO CHAPTER 3

1. Yi Chungnye, Taped Interview, June 3, 1999; Kim Chŏngmin, Taped Interview, June 8, 1999; Kim Ŭnnye, Taped Interview, June 12, 1999.

2. U Sunok, "Yŏgong ilgi" [Diary of a factory girl], *Pyŏlgŏn'gon* (March 1930), 72–73.

3. Michel Foucault, *Discipline and Punish: The Birth of the Prison* (New York: Vintage, 1979), 128–54.

4. Ching Kwan Lee, "Familial Hegemony: Gender and Production Politics on Hong Kong's Electronics Shopfloor," *Gender & Society* 7:4 (December 1993), 531.

5. Michel de Certeau, *The Practice of Everyday Life* (Berkeley: University of California Press, 1984), xix; John Law, *Organizing Modernity* (Oxford: Blackwell, 1994).

6. Patrick Joyce, *The Rule of Freedom: Liberalism and the Modern City* (London: Verso, 2003), 6.

7. Tamara Hareven, *Family Time and Industrial Time: The Relationship Between the Family and Work in a New England Industrial Community* (Cambridge: Cambridge University Press, 1978), 154.

8. Joan W. Scott and Louise A. Tilly, *Women, Work and Family* (New York: Routledge, 1978), 119; Angelique Janssens, *Family and Social Change: The Household as a Process in an Industrializing Community* (Cambridge: Cambridge University Press, 1993), 223–46; Hareven, *Family Time and Industrial Time*, 154.

9. Foucault, *Discipline and Punish*, 83–84, 148. See also Alvin Gouldner, *Patterns of Industrial Bureaucracy: A Case Study of Modern Factory Administration* (New York: Macmillan, 1954).

10. de Certeau, *The Practice of Everyday Life*, 32.

11. During the incubation period, newly hatched larvae feed on chopped mulberry leaves strewn on feeding trays. Because of the quantity of mulberry leaves required by silkworms (approximately 485 pounds of chopped mulberry leaves are necessary to produce two pounds of silk), the management of mulberry fields was as significant as the process of incubation.

12. Silk that is unsuitable for reeling is cleaned, degummed, carded, combed, produced into slivers, and finally spun to make silk yarns. By the 1930s, most silk-reeling factories in Korea had power looms to weave the silk, but workers were still needed to piece together the filaments that frequently broke from tension.

13. Kim P'yŏngun, *Myŏnbangjŏk sŏmyu kongŏp ch'ongsŏ* [Documents concerning the cotton-spinning fiber industry], vol. 1 (Seoul: Ŭlyumunhwasa, 1948), 4–5.

14. When a yarn broke as it issued forth from the machinery, it reverted backwards due to the spinning motion. The piecer had to catch the yarn and tie it to the end of the reeled yarn to continue the reeling process. Although piecers were most prominent in cotton spinning, all divisions of textile production, including cotton weaving as well as silk spinning and weaving, required the work of piecers.

15. According to testimonies, women were assigned to their tasks and divisions after brief physical examinations. Stronger women, for example, were designated to the mixing and scutching (*hont'amyŏn*) rooms, along with men. With the shortage of labor in latter years of the colonial era, even novices began to fulfill the more dangerous and previously male-dominated positions. Yi Chungnye, for example, lost her forearm as it was drawn into a carding machine. This incident clearly indicates that as a peasant girl of fourteen, she was first misplaced into one of the most dangerous sections in the factory, the carding room. Yi Chungnye, Taped Interview, June 3, 1999.

16. It was not until the mid-1920s, with the invention of the Toyoda automatic loom by Toyoda Automatic Loom Works in 1926, that textile production rapidly

advanced in Korea. According to a contemporary report by Arno Pearse, the outstanding merit of the Toyoda loom was the high speed, which was "250 picks per minute." Other inventions existed and the "Kanegafuchi Spinning and Weaving Company had automatic looms of its own invention." Arno S. Pearse, *The Cotton Industry of Japan and China* (Manchester, UK: International Cotton Federation, 1929), 79–83.

17. A *reed* is a narrow movable frame with reed or metal strips that separated the warp threads in weaving.

18. For variety, I refer to knitting and hosiery interchangeably. The four categories of materials that were used for clothing (*ŭibok*) were *chigmul* (woven material), *p'yŏnmul* (knitted material), felt, lace, and net fabric (*mang*), which was an openwork fabric made of threads or cords that were woven or knotted together at regular intervals. The Korean term for knitting (*meriasŭ*) was imported from Japanese knitting industries (*meriasu kōgyō*), which adopted the Portuguese term *meias*, derived from the Latin *maiorinus*.

19. The dyeing industry (*yŏmsaekŏp*) emerged in the 1920s. At the end of 1921, the Tōyō affiliation (*kyeyŏl*) established a specialized dyeing factory in Seoul. Because the dyeing industry was small in scale, Korean businessmen invested in dyeing enterprises soon after the abolition of the Company Law in 1920. In 1924, Paekhwa Yŏmsack (Paekhwa dyes) was established in Pyongyang by Chŏng Insuk, and in 1926, Yi Ch'ang-ho founded Taedong Dyes (Taedong yomsaegŏp) in Taedong. *Tonga ilbo*, September 2, 1922; Heijōfū [Ministry of Pyongyang], *Sangyō chōsushō* [Writings on industry] (Heijō: Heijōfū, 1925), 10–24.

20. Whereas in spinning and weaving, the pace of work was set by machinery, in rubber shoemaking, the speed of the work rested more on the personal decisions of the overseer. Thus, rubber workers strikes arose in response to specific conditions of labor and wage fluctuations rather than industry-wide structures.

21. According to a 1935 study conducted by the Keijōfū [Ministry of Seoul], rubber factories employed an average of approximately 2.7 women workers per male worker. Keijōfū, *Gomu kogyo nigan suru chōsa* [A study of the development of the rubber industry] (Keijō: Keijōfū, 1935), 59–60; "Nebyuhwahan kŭndae saenghwal: Tohoe ka nahŭn kŭnjak ch'il kyŏng" [A review of modern life: The latest work of cities, Part 7], *Sintonga*, 2:6 (June 1932), 6–7.

22. Yuk's research concerned the Chongyŏn (Kanegafuchi) silk-reeling factory in Kwangju. Yuk Chisu, "Seishi rōdō ni tsuite," *Chōsen sōtokufu chōsa geppō* 6:7 (July 1938), 27–40; Chŏng Kŭnsik, "Ilcheha chongyŏn pangjŏk ŭi chamsaŏp chibae" [Labor control of the Chongyŏn (Kanegafuchi) silk-reeling factory under Japanese colonial rule], in *Han'guk kŭndae nongch'on sahoe wa Ilbon cheguk chuŭi* [Rural society and Japanese Imperialism in Modern Korea], Han'guk sahoesa yŏn'guhoe nonmunjip, vol. 2 (Seoul: Munhak kwa chisŏngsa, 1986), 147–94.

23. Pak Ŭnsik, "Ilcheha ŭi yŏsŏng nodong undong: Minjŏk ch'abyŏl ŭi sŏngch'abyŏl kkaji kyŏkŏ" [The female labor movement under Japanese rule: The

experience of ethnic and sex discrimination], *Yŏsŏng*, no. 272 (March 1990), 28–31; Yuk, "Seishi rōdō ni tsuite," 27–33.

24. *Chungwoe ilbo*, June 23, 1929.

25. Although strikes concerning wages were more frequent, many workers appealed for the improvement of tools and machinery, claiming that products made with defective instruments could not meet regulation standards. According to the participants of the Tongsin Silk Reeling strike of mid-February 1931 in Seoul, although the company docked wages because of inferior products, these shortcomings were not due to the workers but the inadequacy of machinery and tools. According to an employee of the Tosi Filature in Kwangju, laborers were continuously suffering from losses due to the large and heavy regulatory penalty system. *Chungwoe ilbo*, June 24, 26, 1929; *Tonga ilbo*, January 13, 1930; *Keijō nippō*, February 16, 17, 1931; Chŏng Chinsŏng, "Ilcheha Chosŏn ŭi ittsŏsŏ nodongja chonjae hyŏngt'ae wa chŏin'gŭm, 1930 nyŏndae rŭl chungsim ŭro" [The living conditions and low wages of Korean workers under Japanese rule: Focusing on the 1930s], in Pak Hyŏnch'ae, ed., *Han'guk chabonjuŭi wa imgŭm nodong* [Wage labor and Korean capitalism] (Seoul: Hwada, 1984); Hŏ Suyŏl, "Ilcheha Han'guk ŭi ittŏsŏ singmingjijŏk sŏnggyŏk e kwanhan ilyŏngu" [The characteristics of colonial industrialization in Korea under Japanese rule], unpublished Ph.D. dissertation, Seoul National University, 1983.

26. Whereas the average Japanese, male factory laborer made 1.87 yen a day, the Korean counterpart earned 0.85 yen. Japanese women received wages comparable to those of Korean men, and Korean women earned around one-half the wages of the Korean men. Chōsen sōtokufu gakumukyoku shakaika, *Kōjō oyobi kōzan ni okeru rōdō jōkyō chōsa* [Survey of labor conditions in factories and mines] (Keijō: Chōsen sōtokufu, 1933), 73–74; Yi Hyojae, "Ilcheha Han'guk yŏsŏng nodong munje yŏn'gu" [The problems of female labor under Japanese rule], in Yun P'yŏngsŏk, Sin Yongha, and An P'yŏngjik, eds., *Han'guk kŭndae saron*, vol. 3 (Seoul: Chisik sanŏpsa, 1977), 97; Kwŏn Yŏnguk, "Nihon teikokushūgika no Chōsen rōdō jijō: 1930 nendai o chūshin ni" [Labor conditions in Chōsen under Japanese imperialism: A focus of the 1930s], *Rekishigaku kenkyū* 303 (August 1965), 25–39.

27. Ch'oe Sukkyŏng, Yi Paeyŏng, Sin Yŏngsuk, and An Yŏnsŏn, "Han'guk yŏsŏngsa chŏngnip ŭl wihan yŏsŏng inmul yuhyŏng yŏn'gu: Samil undong ihu put'ŏ haebang kkaji" [A study of characteristics of notable women for the strengthening of Korean women's history: From the March First movement to liberation], *Yŏsŏnghak nonjip*, vol. 10 (Seoul: Ewha Woman's University and Han'guk yŏsŏng yŏn'gusoe, 1993), 11–129; Andrew J. Grajdanzev, *Modern Korea* (1944; reprint, New York: Octagon Books, 1978), 183–84; Suzuki Masabumi, *Chōsen keizai no gendankai* [The current stage of the Korean economy] (Keijō: Teikoku chihō gyōsei gakkai Chōsen honbu, 1938).

28. Even in factories that housed most of their employees, day labor still existed. But the vast majority of female workers in silk and cotton textile mills lived in com-

pany housing. Of eight former factory workers, Kim Ŭnnye was the only day laborer, as she commuted from home while working at the Pyongyang Hosiery factory from 1936 to 1939, as well as a tobacco factory also in Pyongyang. Kim professes that she awoke "extremely early" because she had to commute to the factory. Details of in-factory mornings were provided by Han Kiyŏng and Kang Pokchŏm, among others. Kang Pokchŏm, Taped Interview, June 3, 1999; Han Kiyŏng, Taped Interview, June 11, 1999; Kim Ŭnnye, Taped Interview, June 12, 1999.

29. Gail Hershatter, *The Workers of Tianjin, 1900–1949* (Stanford, CA: Stanford University Press, 1986), 151; Kim Chongsŏng, *Han'guk ŭi imgŭm mit nodongja ŭi kwanhan ilyŏn'gu: Ilcheha ŭi nodongja sangt'ae rŭl chungsim ŭro* [A study of work, wages, and conditions of Korean workers under Japanese rule] (Pusan: Kyŏnsang taehakkyo inmun sahoe kwahak p'yŏn, 1982).

30. Although in Japan the Safety Regulations for Factories of June 1929 included stipulations for the control of dust, in Korea, no such measures existed. Kwŏn Minjŏng, "Yŏsŏng nodongjae ŭi chikŏpsŏng chilhwan ŭi kwanhan yŏn'gu" [A study of the occupational diseases of women workers], *Yŏsŏng yŏn'gu* 7:2 (Summer 1989), 94–107. Honig cites Xia Yan, who noted that on an average twelve-hour shift, workers in spinning divisions inhaled 0.15 grams of cotton dust. Emily Honig, *Sisters and Strangers: Women in the Shanghai Cotton Mills, 1919–1949* (Stanford, CA: Stanford University Press, 1986), 142.

31. Chōsen sōtokufu gakumukyoku shakaika, *Kōjō oyobi kōzan*, 193–203.

32. For labor strikes among silk workers, see *Tonga ilbo*, July 18, 1931, November 16, 1932, February 24, 1932, September 1, 1932, December 6, 1934, December 26, 1934, September 8, 1938.

33. Chōsen sōtokufu gakumukyoku shakaika, *Kōjō oyobi kōzan*, 49.

34. Han Kiyŏng, Taped Interview, June 11, 1999; Kang Pokchŏm, Taped Interview, June 3, 1999.

35. Honig, *Sisters and Strangers*, 142; Kang Pokchŏm, Taped Interview, June 3, 1999.

36. E. Patricia Tsurumi, *Factory Girls: Women in the Thread Mills of Meiji Japan* (Princeton, NJ: Princeton University Press, 1990), 132. See also Thomas Dublin, *Women at Work: The Transformation of Work and Community in Lowell, Massachusetts, 1826–1860* (New York: Columbia University Press, 1981). Hershatter, *The Workers of Tianjin*; Honig, *Sisters and Strangers*.

37. Whereas in Meiji mills, where the two women sharing the sleeping area sometimes shared the same sleep garments, among the Korean textile workers interviewed, no mention was made of undergarments or sleepwear given by the companies. Such kinds of deficiencies prompted protest, as within the nine demands presented, the workers of Tongyang Chesa factory in Chinhae called for the institution of separate bathrooms and improvement of bathing facilities. *Pusan ilbo*, January 7, 10, 13, February 4, 16, 1933; Tsurumi, *Factory Girls*, 133; Kang Pokchŏm, Taped Interview, June 3, 1999; Pak Sun'gŭm, Taped Interview, June 8, 1999.

38. *Tonga ilbo*, January 13, 1930; Han'guk nodong chohap ch'ongyŏnmaeng [The Amalgamated Labor Unions of Korea], ed., *Han'guk nodong chohap undongsa* [History of Korean labor unions] (Seoul: Han'guk nodong chohap ch'ongyŏnmaeng, 1991), 183.

39. Kang Pokchŏm, Taped Interview, June 3, 1999.

40. Pak Sŏngch'ŏl, ed., *Kyŏngbang ch'ilsimnyŏnsa* [Seventy-year history of Kyŏngbang] (Seoul: Kyŏngsŏng pangjik chusik hoesa, 1989), 114; Cho Yŏnggu, ed., *Kyŏngsŏng pangjik osimnyŏnsa* [Fifty-year history of the Kyŏngsŏng Spinning and Weaving Company] (Seoul: Kyŏngsŏng pangjik chusik hoesa, 1969).

41. Systems of enclosure, partitioning, and discipline, as elaborated by Foucault in his description of creating "docile bodies," were adopted in colonial Korean factories. Foucault, *Discipline and Punish*, 135–69.

42. In most cases, overseers walked about their designated sections of the room and supervised the activities of workers. They generally held small canes and, in some cases, wrenches or other metal objects, to tap the workers for attention. Kang Pokchŏm, Taped Interview, June 3, 1999.

43. Although less common, workers, such as those of the Pyongyang rubber factory in Silli, called for the reinstatement of favored Japanese supervisors during strikes. *Tonga ilbo*, May 20, 1931, November 10–17, 1932, December 6, 1932; *Chungang ilbo*, November 9, 1932, December 6, 1932.

44. *Keijō nippō*, September 10, 12, 1931; *Pusan sinmun*, September 11, 1931; *Minbo*, September 11, 1931; *Honam sinmun*, September 10, 1931.

45. The abolition of night work for women and children was one of the articles of the Factory Law of 1911, revised in 1923. Although implemented in Japan, the strong opposition of employers "won a postponement of implementation of the clause prohibiting night work until 1929." Barbara Molony, "Equality Versus Difference: the Japanese Debate Over Motherhood Protection, 1915–1950, in Janet Hunter, ed., *Japanese Women Working* (London: Routledge, 1993), 124; Hershatter, *The Workers of Tianjin*, 150.

46. In contemporary Japan, the provisions of the Factory Law barred women and minors from night work for reasons of "motherhood protection" because female night workers, who lost up to 638 grams of body weight during an average night shift, often faced problems with childbirth later in life. Authorities believe that the spread of infectious diseases such as tuberculosis, and dietary ailments including beriberi, as well as the heightened risk of sexual assault necessitated limited degrees of women's and children's labor protection. Tsurumi, *Factory Girls*, 169.

47. Yi Chungnye, Taped Interview, June 3, 1999; Carter J. Eckert, *Offspring of Empire: The Koch'ang Kim's and the Colonial Origins of Korean Capitalism, 1876–1945* (Seattle: University of Washington Press, 1991), 194.

48. Yi Chaeyun, Taped Interview conducted by Kang Isu, October 4, 1991.

NOTES TO CHAPTER 4

1. I refer to industrial labor strikes using a variety of terms, including "boycotts," "protests," "demonstrations," and "uprisings." Chōsen sōtokufu keimukyoku, *Saikin ni okeru Chōsen chian jōkyō* [Recent security conditions in Korea] (Keijō: Chōsen sōtokufu, 1933), 143–47.

2. For more on the Yŏnghŭng general strike, see *Tonga ilbo*, November 24, 1928, December 2, 4, 6, 1928. On Wŏnsan, see *Tonga ilbo*, September 19, 1928, January 15, 18, 26, 29, 1929; *Chosŏn ilbo*, January 23, 25, 1929, February 17, 1929; Hamhŭng hyŏnbyŏngdae (The Military Police Precinct of Hamhŭng), *Wŏnsan nodong chaengŭi wa minjokkwangye koch'al* [The reports of the Wŏnsan labor strikes and ethnic relation], Unpublished report (Hamhŭng: Hamhŭng hyŏnbyŏngdae, February 11, 1929); Kim Chun, "Ilcheha nodong undong ŭi panghyang chŏnhwan ŭi kwanhan yŏn'gu" [Changes in the direction of the workers' movement under Japanese rule], in Han'guk sahoesa yŏn'guhoe, ed., *Ilcheha ŭi sahoe undong*, Han'guk sahoesa yŏn'guhoe nonmunjip, vol. 9 (Seoul: Munhak kwa chisŏngsa, 1987), 11–75; Kwŏn Ŭisik, "Uri nara esŏ nodong kyeg'ŭp hyŏngsŏng kwanjŏng kwa kŭ sigi" [The formation of the working class in our nation and the era], *Ryŏksa kwahak* (January 1966), 1–9; Yi Kuksun, "1930 nyŏndae Chosŏn nodong kyegŭp ŭi kusŏnge taehayŏ" [Regarding the formation of the Korean working class in the 1930s], *Ryŏksa kwahak* 1 (April 1963), reprinted in Kim Kyŏngil, ed., *Puk-Han hakkye ŭi nonong undong yŏn'gu* [North Korean studies of labor and peasant movements] (Seoul: Ch'angjak kwa pip'yŏngsa, 1989), 114–36; Hagen Koo, "The State, Minjung and the Working Class in South Korea," in Hagen Koo, ed., *State and Society in Contemporary Korea* (Ithaca, NY: Cornell University Press, 1993), 134; Soon-Won Park, *Colonial Industrialization and Labor in Korea: The Onoda Cement Factory* (Cambridge, MA: Harvard University Press, 1999), 122. For Japanese examples, see Andrew Gordon, *The Evolution of Labor Relations in Japan: Heavy Industry, 1853–1955* (Cambridge, MA: Harvard University Press, 1988), 175–90, 107–21, 246–54.

3. Chōsen sōtokufu, *Chōsen sōtokufu tōkei nenpō* [Statistical yearbook of the Government-General of Korea] (Keijō: Chōsen sōtokufu, 1925, 1926). Statistic for American women cited in Andrew J. Grajdanzev, *Modern Korea* (1944; reprint, New York: Octagon Books, 1978), 184.

4. Victor Turner, *Dramas, Fields, and Metaphors: Symbolic Action in Human Society* (Ithaca, NY: Cornell University Press, 1974), 96.

5. Ching Kwan Lee, "The Unmaking of the Chinese Working Class in the Northeastern Belt," in Ching Kwan Lee, ed., *Working in China: Ethnographies of Labor and Workplace Transformation* (New York: Routledge, 2006), 25.

6. The Rubber Workers' General Strike in Seoul, launched in June 1923, involved Haedong, Noegu, Tongyang, Hansŏng, and Kyŏngsŏng rubber factories (*komu kongjang*). Although rubber production had frequently been seen as a "heavy industry" that

employed male workers, contemporary newspaper reports characterized the "cries of sorrow of female workers in rubber factories" during the strikes. *Chosŏn ilbo*, June 21, 1923, June 28, July 3, 4, 5, 1923; *Tonga ilbo*, July 3, 4, 1923. For examples of women workers' strikes in Taejŏn, see *Tonga ilbo*, April 6, 1929; for Ch'ŏngju, see *Keijō nippō*, September 19, 1931; for Yesan, see *Tonga ilbo*, October 28, 1933.

7. Kwŏn Chungdong, *Yŏja nodong* [Female Labor] (Seoul: Chungang kyŏngjesa, 1986), 37.

8. *Tonga ilbo*, April 16, 1925; Sŏ Hyŏngsil, "Singminji sidae yŏsŏng nodong undong ŭi kwanhan yŏn'gu: 1930 nyŏndae chŏnban'gi komu chep'um chejoŏp kwa chesaŏp ŭl chungsim ŭro" [A study of the female labor movement during the colonial era: A focus on the rubber and silk industries of the early 1930s], in Han'guk chŏngsindae munje taech'aek hyŏpŭihoe and Chŏngsindae yŏnguhoe, eds., *Ilcheha ŭi sahoe undong kwa nongch'on sahoe* [The social movement of the colonial era and rural society], Han'guk sahoesa yŏn'guhoe nonmunjip. vol. 25 (Seoul: Munhak kwa chisŏngsa, 1990), 68–190.

9. Kim Chungyŏl, "Pyongyang komu kongjang p'aŏp" [The strike of the Pyongyang rubber factory], *Nodong kongnon* (March 1975), 107–16; Song Chiyŏng, "1930 nyŏn Pyongyang komu kongjang nodongja dŭl ŭi ch'ŏng p'aŏp" [The general strike of Pyongyang rubber workers in 1930], *Kŭlloja* (May 1959), reprinted in Kim Kyŏngil, ed., *Puk-Han hakkye ŭi nonong undong yŏn'gu* [North Korean studies of labor and peasant movements] (Seoul: Ch'angjak kwa pip'yŏngsa, 1989), 235–53.

10. The 1930 Pyongyang and the 1933 Pusan strikes were prompted by wage cuts, following recession in the rubber industry during the first years of the Depression. On August 1, 1930, members of the Pyongyang Amalgamated Rubber Workers Union (Pyongyang komu chikkong chohap), Pyongyang Federation of Workers (Pyongyang nodong yŏnmaeng), and rubber workers waged a general strike in response to the wage reductions imposed by the Association of Rubber Industrialists of Korea (Chŏn Chosŏn komu kongŏpja taehoe) in May 1930. After 1930, the methods of demonstration among Pyongyang rubber shoemakers diversified, as the employees of several rubber companies went on strike in cycles throughout the early 1930s. *Chosŏn ilbo* August 15, 18, 27, 1930; *Tonga ilbo* September 4, 5, 7, 10, 1930, October 18, 20, 29, 31, 1930, November 12, 1930; Kim Kyŏngil, "Ilcheha komu nodongja ŭi sangt'ae wa nodong undong" [The conditions of rubber workers and the labor movement under Japanese rule], in Han'guk sahoesa yŏn'guhoe, ed., *Ilcheha ŭi sahoe undong*, Han'guk sahoesa yŏn'guhoe nonmunjip, vol. 9 (Seoul: Munhak kwa chisŏngsa, 1987), 76–157; Yi Hyojae, "Ilcheha ŭi yŏsŏng nodong munje" [The problems of female labor under Japanese rule], in An P'yŏngjik and Cho Kijun, eds., *Han'guk nodong munje ŭi kujo* [The structure of the Korean labor movement] (Seoul: Kwangminsa, 1978), 131–79.

11. Kim Chungyŏl, "Chosŏn pangjik kongjang ŭi chaengŭi" [The factory strike of Chōsen Spinning and Weaving], *Nodong kongnon* (January–February 1975), 69–76.

12. Hosiery strikes were prompted by wage cuts resulting from the introduction of circular knitting machines and looms in 1926. Socialist infiltration sometimes contributed to the unrest of the early 1930s. Among others, strike leaders were found to be Yonagawa Akiho, a reporter for the *Sŏsŏn ilbo*, and a student leader, Kim T'aesŏk. *Chungang ilbo*, April 3, 1932. For more on the strikes, see *Chungang ilbo*, March 14, April 3, 1932; *Tonga ilbo*, February 18, March 11, 1932; *Maeil sinbo*, March 15, 16, 17, 1932.

13. *Tonga ilbo*, June 23, 1932.

14. The workers of the Kunsi (Kunze) silk-reeling factory in Taejŏn also sought the reduction of work hours from thirteen to ten hours a day. *Tonga ilbo*, July 31, 1931, August 21, November 10, 17, December 6, 1932; *Chungang ilbo*, November 10, 17, 1932.

15. *Tonga ilbo*, April 6, 1929, October 28, 1933; *Keijō nippō*, September 19, 1931.

16. The Michize silk-reeling factory workers' strike, involving over 600 female workers, lasted from November 9 to 14, 1932. The Kanegafuchi (Chongyŏn) silk-reeling factory workers' strike erupted on November 26, 1932, and involved 158 dormitory workers. *Tonga ilbo*, November 20, 30, 1932; *Pusan ilbo*, November 10, 28, 1932; *Mokp'o ilbo*, November 28, 29, 1932; *Keijō nippō*, August 2, 1932.

17. Han'guk nodong chohap ch'ongyŏnmaeng, ed., *Han'guk nodong chohap undongsa* [History of Korean labor unions] (Seoul: Han'guk nodong chohap ch'ongyŏnmaeng, 1979), 182–209.

18. In newspaper reports, this *kye* among colonial workers was called *ch'inbo kkuhoe*, meaning "friends' savings and loan association." *Keijō nippō*, July 17, 1930, March 14, 1931; *Maeil sinbo*, July 17, 1930; *Chosŏn sinbo*, July 17, 1930; *Pusan ilbo*, March 13, 1931; *Tonga ilbo*, December 25, 1932, February 21, 1933, March 3, 1933; *Chungang ilbo*, December 24, 1932, January 5, 13, 1933; *Pusan ilbo*, January 13, February 11, March 2, 1933.

19. Child workers were fourteen years of age or younger. For more on industrial hierarchy, see An P'yŏngjik, "Singminji Chosŏn ŭi koyong kujo e kwanhan yŏn'gu" [A study of the employment structure in colonial Korea], in An P'yŏngjik, Yi Taegŭn, Nakamura Tetsu, and Kajimura Hideki, eds., *Kŭndae Chosŏn ŭi kyŏngje kujo* [The economic structure of modern Korea] (Seoul: Pibong ch'ulp'ansa, 1989), 388–426.

20. John Rule, "The Property of Skill in the Period of Manufacture," in Patrick Joyce, ed., *The Historical Meanings of Work* (Cambridge: Cambridge University Press, 1987), 99–118.

21. Kang Pokchŏm, Taped Interview, June 3, 1999; Han Kiyŏng, Taped Interview, June 11, 1999.

22. Phillip Ackerman, "Determinants of Individual Differences During Skill Acquisition: Cognitive Abilities and Information Processing," *Journal of Experimental Psychology: General* 117 (1988), 288–318.

23. For more on tactics of resistance, see Pierre Bourdieu, *Outline of a Theory of Practice*, Richard Nice, trans. (Cambridge: Cambridge University Press, 1977), 168;

*The Logic of Practice*, Richard Nice, trans. (Cambridge: Polity Press, 1990); James C. Scott, *Weapons of the Weak: Everyday Forms of Peasant Resistance* (New Haven, CT: Yale University Press, 1985); Elizabeth Janeway, *Powers of the Weak* (New York: Morrow Quill, 1981); Slavoj Žižek, *The Sublime Object of Ideology* (London: Verso, 1989); Ernesto Leclau and Chantel Mouffe, *Hegemony and Social Strategy: Towards a Radical Democratic Politics* (London: Verso, 1985); Judith Butler, Ernesto Leclau, and Slavoj Žižek, *Contingency, Hegemony and Universality: Contemporary Dialogues on the Left* (London: Verso, 2000).

24. Anthony Giddens, *The Constitution of Society: Outline of the Theory of Structuration* (Berkeley: University of California Press, 1984), 14–16, 256–74; *Modernity and Self-Identity: Self and Society in the Late Modern Age* (Stanford, CA: Stanford University Press, 1991), 7; Kenneth Gergen, *The Saturated Self: Dilemmas of Identity in Contemporary Life* (New York: Basic Books, 2000), 8; Judith A. Howard, "From Changing Selves Toward Changing Society," in Judith A. Howard and Peter L. Callero, eds., *The Self-Society Dynamic: Cognition, Emotion and Action* (Cambridge: Cambridge University Press, 1991), 210, 213.

25. According to Kang, most women intended to return home upon the completion of their contracts. Kang Pokchŏm, Taped Interview, June 3, 1999.

26. The Kyŏngsŏng Textile Corporation also had an educational curriculum consisting of two-hour classes in Japanese, arithmetic, and history, as well as vocational information such as silks and satins (*chudan*). In the late 1930s, attempts to institute systems of formal education (*chŏnggyu kyoyuk*) alongside factories commenced. The 1939 opening of the Tonggwang Women's Middle School (Tonggwang chunghakkyo), adjacent to the Kyŏngsŏng Spinning and Weaving Namman factory in southern Manchuria, was such an example. Han Kiyŏng, June 11, 1999. Pak Sŏngch'ŏl, ed., *Kyŏngbang ch'ilsimnyŏnsa* [Seventy-year history of Kyŏngbang] (Seoul: Kyŏngsŏng pangjik chusik hoesa, 1989), 366–67.

27. Cotton smuggled was used as sanitary napkins. Kang Pokchŏm, June 3, 1999.

28. Kang Pokchŏm, June 3, 1999; Han Kiyŏng, June 11, 1999; Bourdieu, *Outline of a Theory*, 39–40; Scott, *Weapons of the Weak*, 29.

29. Yamamoto Shigemi, *Aa Nogmugi tōge* [Ah the Nogumi pass] (Tokyo: Kadokawa, 1977), 97, as cited in E. Patricia Tsurumi, *Factory Girls: Women in the Thread Mills of Meiji Japan* (Princeton, NJ: Princeton University Press, 1990), 90.

30. *Tonga ilbo*, October 12, 1926, March 25, 1931.

31. Kang Pokchŏm recalled that she was made to stand on a platform half-naked from the waist down after a runaway attempt. Nevertheless, it is important to recognize that women in domestic and commercial service were just as vulnerable to violence, rape, and mistreatment, often without pay. Despite their vulnerability, factory girls might have been better off than their rural counterparts inasmuch as they did not endure, or escape from, mistreatment alone. Kang Pokchŏm, June 3, 1999. Yi Chaeyun, October 4, 1991.

32. *Tonga ilbo*, October 27, 1929; February 14, 1936; *Chungang ilbo*, November 21, 1935; June 11, 1936; *Chosŏn ilbo*, September 12, 1930.

33. Kang Pokchŏm, June 3, 1999.

34. Han recalled that in the last months before the war's end in August 1945, over half of her section had run away, indicating that not only productive, but also supervisory, conditions deteriorated considerably by 1945. Han Kiyŏng, June 11, 1999; Kang Pokchŏm, June 3, 1999; Chosŏn nodong chohap chŏn'guk p'yŏngŭihoe ed., *Chŏn'guk nodongja sinmun* (1945; reprint, Seoul: Kukhak charyowŏn, 1998), also reprinted in Han Hŭng-gu, Yi Chae-hwa, eds, *Chosŏn minjok haebang undongsa charyo ch'ongsŏ*, Volume 4 (Seoul: Kyŏngwŏn munhwasa, 1988), 465; Yun Chunha, ed., *Onu yŏsŏng ŭi norae* [Songs of working women] (Seoul: Ingan tosŏ ch'ulp'an, 1983).

35. Lenard L. Berlanstein, ed., *Rethinking Labor History: Essays on Discourse and Class Analysis* (Urbana: University of Illinois Press, 1993), 5.

NOTES TO CHAPTER 5

1. For the purposes of a discussion of wartime Korea, the colonial era can be divided into the early (1910–37) and the late (1937–45) periods. This account is based on an interview with Kang Pokchŏm, Taped Interview, June 3, 1999.

2. Since the Pacific War in East Asia, which opened new fronts in 1931, 1937, and 1941, respectively, was one of the many conflicts that comprised World War II, I use *World War II* interchangeably with *the Pacific War*, particularly in reference to the years of mobilization between 1937 and 1945.

3. John Dower, *Empire and Aftermath: Yoshida Shigeru and the Japanese Experience, 1878–1954* (Cambridge, MA: Harvard University Press, 1979), 85; Louise Young, *Japan's Total Empire* (Berkeley: University of California Press, 1998), 115–80; Kobayashi Hideo, *Daitōa kyōeiken no keisei to hōkai* [The rise and fall of the East Asian Co-Prosperity Sphere] (Tokyo: Ochanomizu Shobō, 1975).

4. Chōsen sōtokufu shokusan kyoku, "Chōsen ni okeru kōjōsu oyobi rōdōshasu," [Statistics on factories and laborers in Korea], *Chōsen sōtokufu chōsa geppō* (January 1941), 10–29; Chōsen sōtokufu, *Chōsen sōtokufu tōkei nenpō* [Statistical yearbook of the Government-General of Korea] (Keijō: Chōsen sōtokufu, 1931, 1939); Andrew J. Grajdanzev, *Modern Korea* (1944; reprint, New York: Octagon Books, 1978), 72–73.

5. Kim Yunhwan, "Kŭndaejŏk imgŭm nodong ŭi hyŏngsŏng kwajŏng" [The formation of modern wage work], in An P'yŏngjik and Cho Kijun eds., *Han'guk nodong munje ŭi kujo* (Seoul: Kwangmingsa, 1980), 59–95; *Han'guk nodong undongsa, Ilcheha p'yŏn* [History of the Korean labor movement under Japanese rule] (Seoul: Ilchogak, 1970); Kim Kyŏngil, *Ilcheha nodong undongsa* [History of labor movements under Japanese rule] (Seoul: Ch'angjakkwa pip'yŏngsa, 1992); *Yi Chaeyu yon'gu: 1930 nyŏndae Seoul ŭi hyŏngmyŏngjŏk nodong undong* [A study of Yi Chaeyu: The revolutionary

labor movement of 1930 Seoul] (Seoul: Ch'angjakkwa pip'yŏngsa, 1993); Yi Hyojae, *Han'guk ŭi yŏsŏng undong: Ŏje wa onŭl* [The Korean women's movement: Yesterday and today] (Seoul: Chŏngwusa, 1989); *Yŏsŏng ŭi sahoe ŭisik* [Women's social consciousness] (Seoul: P'yŏngminsa, 1980); Kang Isu, "Kongjang ch'eje wa nodong kyuyul" [The factory system and labor control], in Kim Chin'gyun and Chŏng Kŭnsik, eds., *Kŭndae chuch'e wa singminji kyuyul kwŏllŏk* [Modernity and colonial disciplinary power] (Seoul: Munhwa kwa haksa, 1997), 117–69; Yi Chŏngok, "Ilcheha kongŏp nodong esŏ ŭi minjok kwa sŏng" [Gender and nationality in industrial labor under Japanese rule], unpublished Ph.D. dissertation, Seoul National University, 1990.

6. The mistranslation of current research organs, namely the Korean Council for Women Drafted for Military Sexual Slavery by Japan (Han'guk chŏngsindae munje taech'aek hyŏpŭihoe) and the Research Association on the Women Drafted for Military Sexual Slavery by Japan (Chŏngsindae yŏn'guhoe) misleadingly indicates that *chŏngsindae* or *teishintai* was synonymous with military sexual slavery. Despite their laudable work in drawing attention to comfort women, the mistranslation of these groups has undermined, although perhaps inadvertently, the history of wartime workers, including the Women's Labor Volunteer Corps. For more on comfort women, see Keith Howard, ed., *True Stories of the Korean Comfort Women* (London: Cassell, 1995), 12; Chunghee Sarah Soh, "From Imperial Gifts to Sex Slaves, Theorizing Symbolic Representations of the 'Comfort Women,'" *Social Science Japan Journal* 3:1 (2000), 59–76; "Human Rights and the 'Comfort Women,'" *Peace Review* 12:1 (January 2000), 123–29; Han'guk chŏngsindae munje taech'aek hyŏpŭihoe [Committee for the resolution of the Korean Women's Volunteer Corps problem] and Chŏngsindae yŏnguhoe [Committee for research on the Korean Women's Volunteer Corps], eds., *Kangjero kkŭllyŏgan Chosŏnin kunwianbudŭl* [The forced migration of Korean comfort women] (Seoul: Hanŭl, 1993); *Chungguk ŭro kkŭllyŏgan Chosŏnin kunwianbudŭl* [The forced migration of Korean comfort women to China] (Seoul: Hanŭl, 1995); Yŏ Sunju, "Ilchemalgi Chosŏnin yŏja kŭllo chŏngsindae ŭi kwanhan sil'tae yŏn'gu" [A study of the realities of the Women's Volunteer Corps in the last years of the colonial era], Unpublished M.A. Thesis, Ewha Woman's University, 1994.

7. *Teishintai* is the Japanese equivalent of *chŏngsindae*.

8. "Mobilization" here and throughout is used to refer to the official recruitment of Koreans for industrial labor. *Maeil sinbo* articles, as well as research from the Rōdōkagaku kenkyūsho [Research Institute of Labor Resource] and Japanese sources such as Kondō Kenichi, ed., *Taiheiyō senka no Chōsen oyobi Taiwan* [Korea and Taiwan during the Pacific War] provide descriptions of volunteers taken to Japan. For more, see Rōdōkagaku kenkyūsho [Research Institute of Labor Resource], "Hantō rōmūsha kinrō jōkyō kansuru chōsa hōkoku" [Report concerning the circumstances of Korean workers], in Pak Kyŏngsik, ed., *Chosŏn munje charyo ch'ongsŏ*,

vol. 1: *Chŏnsi kangje yŏnhaeng: Nomu kwalli chŏngch'aek* [Collection of sources on problems in Korea, vol. 1: The displacement of workers during wartime: Policies on labor management]. vols. 1 and 2 (1943; reprint, Tokyo: Asia mondai kenkyūsho, 1982); Kondō Kenichi, ed., *Taiheiyō senka no Chōsen oyobi Taiwan* [Korea and Taiwan during the Pacific War] (1961; reprint, Seoul: Kukhak charyowŏn, 1984); *Taiheiyō senka shūmatsuki Chōsen no chisei* [Policy in Korea in the last period of the Pacific War] (1961; reprint, Seoul: Kukhak charyowŏn, 1984).

9. From this point on, I refer to the Society for the Survivors and the Bereaved Families of the Pacific War simply as "the society." T'aep'yŏngyang chŏnjaeng hŭisaengja yujokhoe, "Kŭllo chŏngsindae hyŏnhwang" [Present conditions of the Women's Labor Volunteer Corps], Unpublished records, Seoul: Society for the Survivors and the Bereaved Families of the Pacific War, 1–3; T'aep'yŏngyang chŏnjaeng hŭisaengja yujokhoe, "The Issue of Korean Human Rights During and After the Pacific War" (Seoul: Society for the Survivors and the Bereaved Families of the Pacific War, March 1993).

10. Young, *Japan's Total Empire*, 55–182; Soon-Won Park, *Colonial Industrialization and Labor in Korea: The Onoda Cement Factory* (Cambridge, MA: Harvard University Press, 1999), 140–41; Kwak Kŏnhong, *Ilche ŭi nodong chŏngch'aek kwa Chosŏn nodongja, 1938–1945* [Labor policy under Japanese rule and Korean workers] (Seoul: Sinsŏwŏn, 2001).

11. W. Donald Smith, "Beyond 'The Bridge on the River Kwai': Labor Mobilization in the Greater East Asia Co-Prosperity Sphere," *International Labor and Working Class History* 58 (Fall 2000), 222; *Chōsen nenkan*, 1941 [Korea yearbook] (Keijō: Keijō Nipponsha, October 1941).

12. Masahisa Kōji, "Chōsen ni okeru rōdōryoku no ryōteki kōsatsu" [A quantitative investigation of labor power in Korea], *Shokugin chōsa geppo* (January 1941), 37–66; "Hangō no jinteki shigen" [Human resource and the Korean Peninsula], *Kokumin bungaku* (May–June 1942), 59; Pak Kyŏngsik, *Ilbon cheguk chuūi ŭi Chosŏn chibae* [The management of Korea during Japanese imperialism] (Seoul: Ch'ŏnga sinsŏ, 1986), 360–61.

13. An ideological campaign better known as the *kōminka* (*hwangminhwa*) movement was one of the facilitators of war mobilization. A contraction of *kōkoku sinmin ka* (*hwangguk sinminhwa*), meaning literally the "conversion of the people of the empire," the program sought to transform Koreans into the emperor's subjects. Chou Wan-yao, "The Kōminka Movement in Taiwan and Korea: Comparisons and Interpretations," in Peter Duus, Ramon H. Myers, and Mark R. Peattie, eds., *The Japanese Wartime Empire, 1931–1945* (Princeton, NJ: Princeton University Press, 1996), 40–69.

14. Passed by Japanese legislators in March 1938, the law (Imperial Ordinance, Issue Number 316) set in motion policies and programs for war mobilization in Japan and its colonies, including Korea. For more extended analysis of the law, see

Kobayashi Hideo, "1930 nendai zenhanki no Chōsen rōdō undō ni tsuite: Heijō gomu kōjō rōdōsha no zenesuto o chūshin ni shite" [The Korean labor movement of the early 1930s: A focus on the Pyongyang rubber workers' general strike], *Chōsenshi kenkyūkai ronbunshū*, vol. 6 (September 1969), 115–22; also reprinted in Namiki Makoto, et al., *1930 nyŏndae minjok haebang undong* [National liberation movements of the 1930s] (Seoul: Kŏrŭm sinsŏ, 1984), 241–69; Kang Tongjin, *Ilbon kŭndaesa* (Seoul: Hangilsa, 1986); Pak Kyŏngsik, *Ilbon cheguk chuŭi ŭi Chosŏn chibae* [The management of Korea during Japanese imperialism] (Seoul: Ch'ŏnga sinsŏ, 1986).

15. According to a January 1941 report, approximately 36 percent of workers in building and construction were recruited through the Employment Promotion system. Although, as noted by W. Donald Smith, the state made "only limited [immediate] use" of the powers granted by the 1938 National General Mobilization Law, these policies and projects established the foundations for the intensification of total war. Hŏ Suyŏl, "1930 nyŏndae kunsu kongŏphwa chŏngch'aek kwa Ilbon tokchŏm chabon ŭi chinch'ul" [The advancement of Japanese capital and wartime industrial policies in the 1930s], in Ch'a Kibyŏk, ed., *Ilche ŭi Han'guk singmin t'ongch'i* [Japanese rule in colonial Korea] (Seoul: Chŏngŭmsa, 1985), 317; Kobayashi Hideo, *Daitōa kyōeiken no keisei to hōkai*, 277–81; Smith, "Beyond 'The Bridge,'" 219–38.

16. The Korean patriotic associations were extensions of the Japanese patriotic labor corps, which by 1944 mobilized 5.7 million Japanese civilians. Chōsen sōtokufu, *Chōsen ni okeru kokumin seishin sōdōin* [National spiritual mobilization in Korea] (Keijō: Chōsen sōtokufu, 1940), 79; Smith, "Beyond 'The Bridge,'" 221.

17. Masahisa Kōji, "Sangyō tōshi to shite no Keijō no genzai oyobi shorai" [Present and future of Kyŏngsŏng as an industrial city], *Chōsen* 312 (September 1941).

18. *Maeil sinbo*, October 24, 1941; Chōsen sōtokufu, *Chōsen ni okeru kokumin seishin sōdōin* [National spiritual mobilization in Korea] (Keijō: Chōsen sōtokufu, 1940).

19. The Japanese Imperial Rule Association (Taisei yokusankai) was created in September 1940 whereas the Japanese National Registration System was launched in November 1941. According to a January 1941 report, approximately 15 percent of workers in building and construction were recruited through the Patriotic Labor Movement. As Soon-Won Park elucidates, the "patriotic unit" was pervasive through the Korean League for National General Mobilization (Kungmin ch'ongnyŏk ch'ongdonwŏn Chosŏn yŏnmaeng), organized in July 1938. The entire peninsula was organized in a regional, province, city, "*kun-ŭp-myŏng-ri-purak*" system. Like the *tonarikumi* organization in Japan, the patriotic unit was composed of ten households, although as few as seven and as many as twenty existed in practice. By 1940, approximately four hundred thousand patriotic units were organized in Korea, covering over 4 million households. Park asserts that "ninety some percent of all Koreans (estimated as around 4.3 million households in total) were organized around one

pyramid structure." *Maeil sinbo*, August 26, 1944; Thomas R. H. Havens, "Women and War in Japan, 1937–1945," *American Historical Review* 80:4 (October 1975), 919; Park, "Colonial Home Front," 11. Ōkurashō kanrikyoku [Ministry of Finance Management Bureau], *Nihonjin no kaigai katsūdō ni kansuru rekiteki chōsa* [History of the Japanese Overseas Activity, Korean section], vol. 10 (1947; reprint, Seoul: Koryŏ sŏrim, 1985); Hŏ, "1930 nyŏndae kunsu kongŏphwa," 317.

20. "General recruitment of labor" did not necessarily mean that workers drafted under such projects were adults. The Women's Labor Volunteer Corps (Yŏja kŭllo ch'ŏngsindae; Joshi seishin kinrōrei), for example, recruited students under the general category of "female mobilization." Rōdōkagaku kenkyūsho [Research Institute of Labor Resource], "Hantō rōmūsha," 74–96, 136–37.

21. Andrew Gordon, *The Evolution of Labor Relations in Japan: Heavy Industry, 1853–1955* (Cambridge, MA: Harvard University Press, 1988), 299–326.

22. On October 2, 1937, Governor-General Minami Jiro received the imperial oath for the emperor's subjects in Korea. Patriotic units were formed not only in factories but in schools, neighborhoods, farms, and offices. Chōsen sōtokufu, *Chōsen jijō* [The situation in Korea] (Keijō: Chōsen sōtokufu, 1940), 102. See also Pak, *Ilbon cheguk chuŭi*, 385; Chou, "The Kōminka Movement," 43.

23. Park, *Colonial Industrialization*, 158.

24. The majority of these workers were employed in mining and manufacturing, accounting for around 10 percent and 85 percent of the total, respectively. Ōkurashō kanrikyoku, *Nihonjin no kaigai katsūdō*, 71; Pak Kyŏngsik, *Ilbon cheguk chuŭi ŭi Chosŏn chibae* [The management of Korea during Japanese imperialism] (Seoul: Ch'ŏnga sinsŏ, 1986), 360–61; Haruko Taya Cook and Theodore Cook, eds., *Japan at War: An Oral History* (New York: New Press, 1992).

25. Through government recommendation, recruitment, and regional mobilization, more than 100,000 workers organized for work in war-related industries in 1938. This figure doubled by 1939 and tripled by 1941. Chōsen sōtokufu, *Chōsen sōtokufu tōkei nenpō* [Statistical yearbook of the Government-General of Korea] (Keijō: Chōsen sōtokufu, 1930, 1935, 1940); Pak Kyŏngsik, *Ilbon cheguk chuŭi*, 360–61; Soon-Won Park, *Colonial Industrialization and Labor*, 139–49.

26. Nearly 3.7 million Koreans worked and resided in Japan and Manchuria by 1945. Macrostatistics for wartime labor remain inconsistent because of the hasty collection and destruction of records. According to Pak Kyŏngsik, however, roughly 7 to 7.5 million Korean workers, soldiers, and military personnel were enlisted for the war effort. Approximately 2.5 million of these workers were taken to Japan, whereas 4.5 million remained in Korea. Pak Kyŏngsik, *Ilbon cheguk chuŭi*, 360–61; Park, "Colonial Home Front," 11; Young, *Japan's Total Empire*, 55–182.

27. E. Patricia Tsurumi, "Colonial Education in Korea and Taiwan," in Ramon H. Myers and Mark R. Peattie, *The Japanese Colonial Empire, 1895–1945* (Princeton, NJ: Princeton University Press, 1984), 305.

28. Kim Chǒngmin, June 8, 1999.

29. Following the Task Concerning the Duties of Labor Leagues of 1939, approximately nine thousand male students from twenty-eight schools and more than four thousand students from nineteen schools assembled for one-week periods of service. *Tonga ilbo*, June 23, 1938.

30. *Maeil sinbo*, April 6, 1944.

31. *Maeil sinbo*, May 9, 13, 1944.

32. Students wrote encouraging messages such as "Do away with rats" for imperial troops. *Tonga ilbo*, June 23, 1938.

33. *Maeil sinbo*, November 25, 1943; Chōsen kōsei kyōkai [Chōsen welfare society], *Chōsen ni okeru jinkō ni kansuru shotōkei* [Statistics on population in Korea] (Keijō: Chōsen kōsei kyōkai, 1945).

34. Yu Kwangnyǒl, "Kyǒlchǒn kungnae t'aese ǔi kanghwa" [The strengthening of wartime national conditions], *Chogwang* (November 1943), 29.

35. *Maeil sinbo*, October 7, 8, 1943 (italics mine).

36. Tsurumi, "Colonial Education," 305.

37. Grajdanzev, *Modern Korea*, 266.

38. Newspaper advertisements of domestic servants entering the corps encouraged other working women to commence patriotic and more profitable work. *Maeil sinbo*, June 29, 30, 1944; October 6, 1944.

39. *Maeil sinbo*, August 26, 1944.

40. Company personnel, accompanied by Korean policemen, recruited Kang Pokchǒm, Yi Chungnye, and Kim Ǔnnye. Yi Chungnye, June 3, 1999; Kang Pokchǒm, June 3, 1999; Kim Ǔnnye, June 12, 1999.

41. *Maeil sinbo*, October 6, 1944.

42. Novices were taken to factories such as the Kanegafuchi factory in Kwangju, the Chōsen factory in Pusan, and the Tōyō, Kyǒngsǒng, and Dai Nippon factories in Yǒngdǔngp'o.

43. *Maeil sinbo*, March 14, 1944.

44. T'aep'yǒngyang chǒnjaeng hǔisaengja yujokhoe, "Kǔllo chǒngsindae hyǒnhwang" [Present conditions of the Women's Volunteer Corps], Unpublished records (Seoul: Society for the Survivors and the Bereaved Families of the Pacific War, 1999), 1–3.

45. Three of the nine informants, Kim Chǒngnam, Kim Chǒngmin, and Pak Sun'gǔm, were employed at the Fujikoshi factory in Toyama.

46. Fujikoshi kōzai kōgyō kabushiki kaisha, *Fujikoshi nijū nenshi* [Twenty-year history of Fujikoshi] (Toyama: Fujikoshi kōzai kōgyō kabushiki kaisha, 1953), 34; *Fujikoshi gojū nenshi* [Fifty-year history of Fujikoshi] (Toyama: Fujikoshi kōzai kōgyō kabushiki kaisha, 1978).

47. Kim Chǒngmin, June 8, 1999.

48. Kim Chǒngnam, June 5, 1999.

49. Fujikoshi kōzai kōgyō kabushiki kaisha, *Fujikoshi nijū nenshi*, 34–44.

50. Han Kiyǒng, June 11, 1999.

51. *Maeil sinbo*, October 28, 1944.

52. Havens, "Women and War in Japan, 1937–1945," 913; Gordon Wright, *The Ordeal of Total War, 1939–1945* (New York: Harper & Row, 1968), 244.

NOTES TO CONCLUSIONS

1. Chǒn died at the age of twenty-three. Chǒn T'aeil, *Nae chugǔm ǔl hǒttoei malla* [Don't waste my life: A collection of essays] (Seoul: Tolbegae, 1988); Kim Nakchung, *Han'guk nodong undongsa, haebanghu p'yǒn* [History of the Korean labor movement after Liberation] (Seoul: Ilchogak, 1970); Seung-Kyung Kim, *Class Struggle or Family Struggle? The Lives of Women Factory Workers in South Korea* (Cambridge: Cambridge University Press, 1997), 100, 132; Hagen Koo, "The State, Minjung and the Working Class in South Korea," in *State and Society in Contemporary Korea* (Ithaca, NY: Cornell University Press, 1993), 131–62; Jang Jip Choi, *Labor and the Authoritarian State: Labor Unions in South Korean Manufacturing Industries, 1961–1980* (Seoul: Korea University Press, 1989); "Political Cleavages in South Korea," in Hagen Koo, ed., *State and Society in Contemporary Korea* (Ithaca, NY: Cornell University Press, 1993), 13–50; Namhee Lee, *The Making of Minjung: Democracy and the Politics of Representation in South Korea* (Ithaca, NY: Cornell University Press, 2007). On Kang Churyǒng, see *Chosǒn ilbo*, May 29–31, 1931; Theodore Jun Yoo, *The Politics of Gender in Colonial Korea: Education, Labor, and Health, 1910–1945* (Berkeley: University of California Press, 2008), 127–29.

2. Han'guk nodong chohap ch'ongyǒnmaeng, ed., *Han'guk nodong chohap undongsa* [History of Korean labor unions] (Seoul: Han'guk nodong chohap ch'ongyǒnmaeng, 1979), 787; *Yǒsǒng nodong kwa p'yǒngdǔng* [Equality and female labor] (Seoul: Han'guk nodong chohap ch'ongyǒnmaeng, 1991).

3. In 1971, there were 101 strikes with 115,934 participants. In spite of the fact that the numbers of participants far surpassed the strikes of the 1930s, the numbers of labor demonstrations never rose as high as the 1930s, implying that greater union organization affected the strikes of the 1970s. Han'guk nodong chohap ch'ongyǒnmaeng, ed., *Han'guk nodong chohap undongsa*, 793; Hwang Sǒkyǒng, "Kuro kongdan ǔi nodong silt'ae" [The realities of labor in the Kuro industrial area], *Wǒlgan chungang* (December 1973), 117–27.

4. Chosǒn nodong chohap chǒn'guk p'yǒngǔihoe ed., *Chǒn'guk nodongja sinmun* (1945; reprint, Seoul: Kukhak charyowǒn, 1998); Charles Armstrong, *The North Korean Revolution: 1945–1950* (Ithaca, NY: Cornell University Press, 2003), 87–93; Soon-Won Park, *Colonial Industrialization and Labor in Korea: The Onoda Cement Factory* (Cambridge, MA: Harvard University Press, 1999), 171.

5. Armstrong, *The North Korean Revolution*, 88, 93.

6. Kong Cheuk, *1950 nyǒndae Han'guk ǔi chabonka yǒn'gu* [History of Korean entrepreneurs in the 1950s] (Seoul: Haksul chongsǒ, 1993), 110; Taehan pangjik hyǒphoe

(Korean Textile Association), *Panghyŏp ch'angnip sipchunyŏn* [The tenth anniversary of the Textile Association] (Seoul: Taehan pangjik hyŏphoe, 1957).

7. Park, *Colonial Industrialization*, 172–77.

8. Although curtailed by the restrictive Yushin Reforms of 1974, workers, including women in light industries, protested with greater vigor by the late 1970s. These strikes began in 1972, subsiding in the late 1980s. Kim Suhwan, ed., *Tongil pangjik nodong chohap undongsa* [The history of the amalgamated workers of the Tongil Spinning and Weaving Company] (Seoul: Tolbegae, 1985), 281–382; Tongil pangjik chusik hoesa, *Tongil pangjik sasa, 1955–1981* [History of the Tongil Spinning and Weaving Company] (Seoul: Tongil pangjik chusik hoesa, 1982); Sun Chŏmsun, *Yŏdŏl sigan nodong ŭl wihayŏ: Hatae chegwa yŏsŏng nodongja dŭl ŭi t'ujaeng kirok* [For eight hours of work: The records of female workers in the Hatai confectionery company] (Seoul: P'ulbit, 1984).

9. *Pusan ilbo*, February 4, 16, 1932.

10. T'aep'yŏngyang chŏnjaeng hŭisaengja yujokhoe [Society for the Survivors and the Bereaved Families of the Pacific War] was founded in April 1973. Kim Ilmyŏn, *Tennō no guntai Chōsenjin ianfu* [Emperor's forces and Korean comfort women] (Tokyo: Sanchi shobo, 1976); T'aep'yŏngyang chŏnjaeng hŭisaengja yujokhoe, "The Issue of Korean Human Rights During and After the Pacific War" (Seoul: Society for the Survivors and the Bereaved Families of the Pacific War, March 1993), 2.

11. Erving Goffman, *Presentation of Self in Everyday Life* (New York: Doubleday, 1959), 17; Edward M. Bruner, "Introduction," and Clifford Geertz, "Making Experience, Authoring Selves," in Victor Turner and Edward M. Bruner, eds., *The Anthropology of Experience* (Urbana: University of Illinois Press, 1986), 22, 378. On roles, see Judith A. Howard and Peter Callero, eds., *The Self-Society Dynamic: Cognition, Emotion and Action* (Cambridge: Cambridge University Press, 1991), 3.

12. Nikolas Rose, *Governing the Soul: The Shaping of the Private Self* (London: Free Association Books, 1999), 103–4.

13. Rose, *Governing the Soul*, 103–4; James Bernard Murphy, *The Moral Economy of Labor: Aristotelian Themes in Economic Theory* (New Haven, CT: Yale University Press, 1993), 225.

14. Viktor Gecas, "The Self-Concept as a Basis for a Theory of Motivation," in Judith A. Howard and Peter Callero, eds., *The Self-Society Dynamic*, 171–88; Jack Barbalet, ed., *Emotions and Sociology* (London: Blackwell, 2002).

15. For more on resistance movements, see Joan Scott, *The Glassworkers of Carmaux: French Craftsmen and Political Action in a Nineteenth-Century City* (Cambridge, MA: Harvard University Press, 1974); Antonio Negri, *Insurgencies: Constituent Power and the Modern State*, Maurizia Boscagli, trans. (Minneapolis: University of Minnesota Press, 1999); Stephen Duncombe, ed., *Cultural Resistance Reader* (New York: Verso, 2002); Richard J. Fox and Orin Starn, eds., *Between Resistance and Revolution* (New Brunswick, NJ: Rutgers University Press, 1997).

16. Louis Althusser, "Ideology and Ideological State Apparatuses," in Hazard Adams and Leroy Searle, eds., *Critical Theory Since 1965* (Tallahassee: Florida State University Press, 1986), 239–51.

17. Rather than the superego, the ego, and the id, Giddens refers to consciousness as discursive, practical, and "unconscious." Whereas discursive consciousness relies predominantly on rationalization, and practical consciousness on stimulation, each of the three levels of consciousness shares mutual influence in the formation of the others. Anthony Giddens, *The Constitution of Society: Outline of the Theory of Structuration* (Berkeley: University of California Press, 1984), 3–37, 64.

18. Judith Butler, *The Psychic Life of Power: Theories in Subjection* (Stanford, CA: Stanford University Press, 1997), 198; Giddens, *The Constitution of Society*, 14–16, 256–74. Rather than the superego, the ego, and the id, Giddens refers to consciousness as discursive, practical, and unconscious. Whereas discursive consciousness relies on predominantly rationalization and practical consciousness on stimulation, each of the three levels of consciousness shares mutual influence in the formation of the others. Giddens, *The Constitution of Society*, 3–37, 64.

19. Gail Hershatter, *The Workers of Tianjin, 1900–1949* (Stanford, CA: Stanford University Press, 1986), 4.

20. Thomas R. H. Havens, "Women and War in Japan, 1937–1945," *American Historical Review* 80:4 (October 1975), 913.

21. Yi Chungnye, June 3, 1999; Yi Chaeyun, Interview conducted by Kang Isu, October 4, 1991.

22. Han Kiyŏng, June 11, 1999; Kim Chŏngnam, June 5, 1999; Kim Chŏngmin, June 8, 1999; Pak Sun'gŭm, June 8, 1999.

23. Barrington Moore, *Injustice: The Social Bases of Obedience and Revolt* (New York: M. E. Sharpe, 1978), 31.

24. Mikhail Bakhtin, *Rabelais and His World*, Helene Isowolsky, trans. (Bloomington: Indiana University Press, 1988), 4–11; Judith A. Howard, "From Changing Selves Toward Changing Society," in Judith A. Howard and Peter L. Callero, eds., *The Self-Society Dynamic*, 224; Paul Redding, *The Logic of Affect* (Ithaca, NY: Cornell University Press, 1999); Thomas J. Scheff, *Microsociology: Discourse, Emotion and Social Structure* (Chicago: University of Chicago Press, 1990); Simon Williams, *Emotion and Social Theory: Corporeal Reflections on the Irrational* (London: Sage, 2001); Giddens, *The Constitution of Society*, 11.

25. Nations industrializing in the nineteenth century used recruitment systems. In contrast to Korea, factory towns functioned in communitarian manners where social activities corresponded to work schedules and descendents carried on their parents' work for two to three generations. Community life often helped the family endure modern transitions. Tamara Hareven, *Family Time and Industrial Time: The Relationship Between the Family and Work in a New England Industrial Community* (Cambridge: Cambridge University Press, 1978), 363–69.

26. Hareven, *Family Time and Industrial Time*, 222; *Families, History and Social Change: Life-Course and Cross-Cultural Perspectives* (Boulder, CO: Westview, 2000).

27. Karl Marx and Friedrich Engels, *The German Ideology*, vol. 1 (New York: International Publishers, 1970), Part 1: Feuerbach: Opposition of the Materialist and Idealist Outlook, 8; *The Communist Manifesto* (1888; reprint, London: Penguin, 1967), 79–90; Karl Marx, *Capital*, vol. 1 (London: Penguin, 1976), 317–19, 54–59; Karl Kautsky, *The Social Revolution* (London: Twentieth Century Press, 1902); Helmut Gruber and Pamela Graves, eds., *Women and Socialism, Socialism and Women* (Oxford, UK: Berghahn Books, 1998), 417–18.

28. Everett Hughes, *The Sociological Eye: Selected Papers*, vol. 1 (Chicago: Aldine-Atherton, 1971), 124.

29. Although they are called "students" (*haksaeng*), they can be better understood as part-time students and full-time workers. Most factory compounds in large cities including Seoul and Pusan no longer have dormitories, but mills in more remote areas, such as Yongin, still recruit and house their workers. Although workers are compensated with overtime wages, night work continues. Articles 62–75, Enforcement Decree of the Act Concerning the Promotion of Worker Participation and Cooperation, Presidential Decree No. 15323, March 27, 1997; Article 65, Enforcement Decree of the Act Concerning the Promotion of Worker Participation and Cooperation, Presidential Decree No. 15323, March 27, 1997; Article 55, Labor Standards Act, Act No. 5309, March 13, 1997. Ministry of Labor, Republic of Korea, *Labor Laws of the Republic of Korea 1997* (Seoul: Munwŏn, 1998); Ministry of Labor, Republic of Korea, *Labor Laws of the Republic of Korea 2005*, http://www.koilaf.org/admin/publication/file/CLS2005.PDF, 13.

30. Kim Ŭnnye, June 12, 1999.

31. Han Kiyŏng, who worked at the Dai Nippon Textile Company in Yŏngdŭngp'o from December 1943 to August 1945, was the only informant who did not work outside of the home after she left the factory. Interestingly, however, Han, like Kim Chŏngnam and Kim Chŏngmin, remembered her wartime labor with sentiments of contribution and accomplishment. Kim Chŏngnam, June 5, 1999; Kim Chŏngmin, June 8, 1999.

32. Yi Chungnye, June 3, 1999.

33. Born in 1925, Kang was sixteen years of age when she entered the factory. From that point on she has been physically, emotionally, and financially independent. Kang Pokchŏm, June 3, 1999.

34. Patrick Joyce, *The Rule of Freedom: Liberalism and the Modern City* (London: Verso, 2003), 7–8.

35. T'aep'yŏngyang chŏnjaeng hŭisaengja yujokhoe, "Kŭllo chŏngsindae hyŏnhwang" [Present conditions of the Women's Volunteer Corps], Unpublished records (Seoul: Society for the Bereaved Families of the Pacific War, 1999); Kang Pokchŏm, June 3, 1999; Kim Ŭnnye, June 12, 1999; Yi Chungnye, June 3, 1999.

36. Jacques Derrida, *Politics of Friendship*, George Collins, trans. (London: Verso, 1997), 277; Jacques Rancière, *The Nights of Labor: The Workers' Dream in Nineteenth-Century France*, John Drury, trans. (Philadephia: Temple University Press, 1989).

37. Maureen Honey, *Creating Rosie the Riveter: Class, Gender and Propaganda During World War II* (Amherst: Massachusetts University Press, 1984), 1.

# Bibliography

PRIMARY SOURCES

An, Tonghyŏk. "Kongŏp Chosŏn ŭi chŏnmo" [The comprehensive portrait of industrialization in Korea]. *Sintonga* (June 1935): 33–46.

Cameron, C. R. "Rubber Industry and Trade of Japan." *Trade Information Bulletin*, no. 354. Washington, DC: U.S. Department of Commerce, Bureau of Foreign and Domestic Commerce, Library of Congress, September 28, 1925.

Ch'a, Kibyŏk, ed. *Ilche ŭi Han'guk singmin t'ongch'i* [Japanese rule in colonial Korea]. Seoul: Chŏngŭmsa, 1985.

Chang, Lin (pseud.). "Nodong puin ŭi chojik hwarŭl" [Concerning strategies for working women]. *Kŭnu* (May 1929).

Chŏk, Ku (pseud.). "Yŏjikkong woesap'yŏn" [Women workers]. *Kaebyŏk* 18:4 (April 1926): 110–11.

Chŏk, Tosaeng. "Nongch'on yŏsŏng pit'ongsa ŭi taehwa" [On the sad history of rural women]. *Chosŏn nongmin* 3:9 (1927).

Chŏng, Osŏng. "Yisip ch'ilnyŏn kan kŭllo hadŭn yŏjikkong An Ahyŏng ŭi adhwa: kŭka kongjang e ssŭlŏjŏ chukki kkaji" [Seventeen years of work, the sad tale of An Ahyŏng] *Sinyŏsŏng* 7:6 (July 1929): 123–43.

"Chŏngch'ijŏk kyŏlsa wa undong e taehayŏ" [On political associations and movements]. *Tonga ilbo*, January 29, 1924.

*Ch'ŏngnyŏn* [Youth].

Chōsen kōgyō kyōkai [Chōsen industrial association]. *Chōsen no kōgyō to sono shigen* [Industry and resources of Chōsen]. Keijō: Chōsen kōgyō kyōkai, 1937.

Chōsen kōsei kyōkai [Chōsen welfare society]. *Chōsen ni okeru jinkō ni kansuru shotōkei* [Statistics on population in Korea]. Keijō: Chōsen kōsei kyōkai, 1945.

*Chōsen nenkan, 1941* [Korea yearbook]. Keijō: Keijō Nipponsha, October 1941.

Chōsen sōtokufu. *Chosen jijō* [The situation in Korea]. Keijō: Chōsen sōtokufu, 1940.

———. *Chōsen kokusei chōsa hōkoku* [National census of the Government-General of Korea]. Keijō: Chōsen sōtokufu, 1925, 1930, 1935, 1940, 1944.

————. *Chōsen ni okeru jinkō kansuru shotōkei* [Statistics on the population of Korea]. Keijō: Chōsen sōtokufu, 1941.

————. *Chōsen ni okeru kokumin seishin sōdōin* [National spiritual mobilization in Korea]. Keijō: Chōsen sōtokufu, 1940.

————. *Chōsen no bussan* [Commodities of Korea]. Keijō: Chōsen sōtokufu, 1927.

————. *Chōsen no keizai jijō* [Economic conditions in Korea]. Keijō: Chōsen sōtokufu, 1931, 1934.

————. *Chōsen no sangyō* [Industries of Korea]. Keijō: Chōsen sōtokufu, 1933, 1935.

————. *Chōsen sembaishi* [Industrial history of Korea]. Vols. 1–3. Keijō: Chōsen sōtokufu, 1936.

————. *Chōsen sōtokufu chōsa geppō* [Monthly report of the Government-General of Korea]. Keijō: Chōsen sōtokufu, 1926–41.

————. *Chōsen sōtokufu tōkei nenpō* [Statistical yearbook of the Government-General of Korea]. Keijō: Chōsen sōtokufu, 1910–45.

————. *Keijō sangkōgyō chōsa* [Industries in Kyŏngsŏng]. Keijō: Chōsen sōtokufu, 1913.

Chōsen sōtokufu gakumukyoku shakaika. *Kaisha oyobi kōjō ni okeru rōdōsha chōsa* [Survey of labor conditions in factories and mines]. Keijō: Chōsen sōtokufu, 1923.

————. *Kōjō oyobi kōzan ni okeru rōdō jōkyō chōsa* [Survey of labor conditions in factories and mines]. Keijō: Chōsen sōtokufu, 1933.

Chōsen sōtokufu keimukyoku. *Saikin ni okeru Chōsen chian jōkyō* [Recent security conditions in Korea]. Keijō: Chōsen sōtokufu, 1933.

Chōsen sōtokufu shokusan kyoku. "Chōsen ni okeru kōjōsu oyobi rōdōshasu" [Statistics on factories and laborers in Korea]. *Chōsen sōtokufu chōsa geppō* [Monthly report of the Government-General of Korea] (January 1941): 10–29.

*Chosŏn chungang ilbo* [Korea Central Daily].

*Chosŏn ilbo* [Korea Daily].

Chosŏn nodong chohap chŏn'guk p'yŏngŭihoe, ed. *Chŏn'guk nodongja sinmun* [National Workers' Newspaper]. 1945. Reprint, Seoul: Kukhak charyowŏn, 1998. Also reprinted in Han Hŭng-gu and Yi Chae-hwa, eds. *Chosŏn minjok haebang undongsa charyo ch'ongsŏ*. Vol. 4. Seoul: Kyŏngwŏn munhwasa, 1988.

*Chosŏn sinbo* [Korea News].

Chu, Ch'ŏn. "Yakhan yŏsŏng kwa nodong kyegŭp ŭi kiwŏn" [The frailty of women and the origin of the working class]. *Sinyŏsŏng* 4:7 (July 1926): 36–37.

*Chungang ilbo* [Central News Daily].

*Chungwoe ilbo* [International Daily].

Hamhŭng hyŏnbyŏngdae [The Military Police Precinct of Hamhŭng]. *Wŏnsan nodong chaengŭi wa minjokkwangye koch'al* [The reports of the Wŏnsan labor strikes and ethnic relation]. Unpublished report. Hamhŭng: Hamhŭng hyŏnbyŏngdae, February 11, 1929.

Han, Kiyŏng. Taped Interview, June 11, 1999.

Heijōfū [Ministry of Pyongyang]. *Sangyō chōsashō* [Writings on industry]. Heijō: Heijōfū, 1925: 10–24.

*Honam sinmun* [Kōnan shimbun, Honam News].

Hulbert, Homer B. "Bible Women." *Korea Review* 6:4 (April 1906): 140–47.

*Kaebyŏk* [Creation].

Kang, Munsŏk. "Chosŏn minjok haebang undong ŭi yissŏ sŏŭi Chosŏn sanŏp nodong chosaso ŭi samyŏng" [A call of the Chosŏn industrial labor research institute]. *Kaebyŏk* 22:1 (January 1930): 59–63.

Kang, Pokchŏm. Taped Interview, June 3, 1999.

Keijōfū. *Gomu kogyo nigan suru chōsa* [A study of the development of the rubber industry]. Keijō: Keijōfū, 1935.

*Keijō nippō* [Kyŏngsŏng Daily].

Keijō shōkō kaigisho [Kyŏngsŏng chamber of commerce and industry]. *Keijō shōko kaigisho tokei nenpō* [The annual statistical report of the Kyŏngsŏng chamber of commerce and industry]. Keijō: Keijō shōkō kaigisho, 1930, 1935.

Kim, Chŏngmin. Taped Interview, June 8, 1999.

Kim, Chŏngnam. Taped Interview, June 5, 1999.

Kim, Myŏngho. "Chosŏn yŏsŏng kwa chigŏp" [Korean women and professions]. *Sinyŏsŏng* 3:10 (October 1925): 9–12.

Kim, Namch'ŏn. "Yŏsŏng ŭi chigŏp munje" [Women's professional problems]. *Yŏsŏng* 5:12 (1940): 22–27.

Kim, Ŭnnye. Taped Interview, June 12, 1999.

Kim, Wŏnju. "Yeppŭni nŭn ŏdiro?" [Where do the attractive go?] *Sinyŏsŏng* 9:12 (December 1931): 105–6.

Kim, Yŏngsŏn. Taped Interview conducted by Kang Isu, June 27, 1991.

Kun, Wŏn (pseud.). "Puin munje ŭi chongjŏn" [Various women's questions]. *Sinsaenghwal* 1:1 (March 1922).

*Kŭnu* [Friends of the Rose of Sharon].

*Kyŏngsŏng pangjik chusik hoesa chonp'yo* [Records of transactions of the Kyŏngsŏng Spinning and Weaving Company]. Unpublished materials. Yongin: Kyŏngsŏng pangjik chusik hoesa ch'amgo, 1940–44.

*Kyŏngsŏng pangjik chusik hoesa chuju ch'onghoerok* [Records of the resolutions of the Board of Directors of the Kyŏngsŏng Spinning and Weaving Company]. Unpublished materials. Yongin: Kyŏngsŏng pangjik chusik hoesa ch'amgo, 1939–47.

*Kyŏngsŏng pangjik chusik hoesa kyŏlŭirok* [Records of the resolutions of the Board of Directors of the Kyŏngsŏng Spinning and Weaving Company]. Unpublished materials. Yongin: Kyŏngsŏng pangjik chusik hoesa ch'amgo, 1938–51.

"Kyŏngsŏng yŏjikkong t'onggye" [Statistics of women workers in Kyŏngsŏng]. *Chosŏn chigwang* 78 (May 1928): 51–52.

*Maeil sinbo* [Mainichi shinpō Daily News].

Masahisa, Kōji. "Chōsen ni okeru chūyō kōjō no rōdō jijō" [Labor situation at the major factories and mines in Korea]. *Shokusan chōsa geppō* [Monthly report on industrial production] 50 (July 1942).

———. "Chōsen ni okeru rōdōryoku no ryōteki kōsatsu" [A quantitative investigation of labor power in Korea]. *Shokugin chōsa geppō* [Monthly report of the industrial bank] (January 1941): 37–66.

———. "Hangō no jinteki shigen" [Human resource and the Korean Peninsula]. *Kokumin bungaku* [National Culture] (May–June 1942): 59.

———. "Sangyō tōshi to shite no Keijō no genzai oyobi shorai" [Present and future of Kyŏngsŏng as an industrial city]. *Chōsen* 312 (September 1941).

*Minbo* [People's News].

Miya, Kōichi. "Chōsen no kairō undo" [The all-labor movement in Korea]. *Chōsen* 318 (November 1941).

———. *Chōsen no rōdōsha* [The laborers of Korea]. Tokyo: Totsu shoseki, 1945.

*Mokp'o ilbo* [Mokp'o Daily].

"Nebyuhwahan kŭndae saenghwal: Tohoe ka nahŭn kŭnjak ch'il kyŏng" [A review of modern life: The latest work of cities, Part 7]. *Sintonga* 2:6 (June 1932).

Ōkurashō kanrikyoku [Ministry of Finance Management Bureau]. *Nihonjin no kaigai katsūdō ni kansuru rekiteki chōsa.* Vol. 10. *Chōsenhen* [History of the Japanese overseas activity, Korean section]. 1947. Reprint, Seoul: Koryŏ sŏrim, 1985.

Pak, Aa. "Yŏsŏng konghwang sidae" [Age of alarm for women]. *Pyŏlgŏn'gon* (July 1932).

Pak, Hojin. "Yŏjikkong pangmungi" [Survey of women workers]. *Kŭnu*, no. 1 (May 1929): 70–73.

Pak, Sun'gŭm. Taped Interview, June 8, 1999.

Pearse, Arno S. *The Cotton Industry of Japan and China.* Manchester, UK: International Cotton Federation, 1929.

*Pusan ilbo* [Fusan nippō, Pusan Daily].

*Pusan sinmun* [Fusan shimbun, Pusan News].

Pyŏl, Moe (pseud.). "Nojajŏn ŭi nalyoe ŭi imhan chungsan kyekŭp ŭi changnae" [Concerning the divide between labor and capital: The future of the middle class]. *Kaebyŏk* 18:2 (February 1926): 19–26.

*Pyŏlgŏn'gon* [Between Heaven and Earth].

Rōdōkagaku kenkyūsho [Research Institute of Labor Resource]. "Hantō rōmūsha kinrō jōkyō kansuru chōsa hōkoku" [Report concerning the circumstances of Korean workers]. In Pak Kyŏngsik, ed., *Chosŏn munje charyo ch'ongsŏ.* Vol. 1: *Chŏnsi kangje yŏnhaeng: Nomu kwalli chŏngch'aek* [Collection of sources on problems in Korea. Vol. 1: The displacement of workers during wartime: Policies on labor management]. Vols. 1 and 2. 1943. Reprint, Tokyo: Asia mondai kenkyūsho, 1982.

Scranton, Mary F. "Woman's Work in Korea." *The Korean Repository* 5:9 (September 1898): 313–18.

*Sin'gajŏng* [New Family].

*Sinmun chŏlbal* [Scrapbooks of newspaper clippings]. Unpublished records. Seoul: Kugwansŏ tosŏgwan [Archive of former government records], Seoul National University Central Libary, Special Collections, 1926–45.

*Sinsaenghwal* [New Life].

*Sintonga* [New East Asia].

*Sinyŏsŏng* [New Woman].

Soe, Mi (pseud.). "Nodongja kyori mundap" [Discussions with workers]. *Kaebyŏk* 18:3 (March 1926): 111–12.

*Sŏjo ilbo* [Western Korea Daily].

Song, Kyewŏl. "Yŏjik kongp'yŏng: kongjang sosik" [Edition on women workers: Factory news]. *Sinyŏsŏng* 9:12 (December 1931), 107–10.

Suzuki, Masabumi. *Chōsen keizai no gendankai* [The current stage of the Korean economy]. Keijō: Teikoku chihō gyōsei gakkai Chōsen honbu, 1938.

T'aep'yŏngyang chŏnjaeng hŭisaengja yujokhoe, "The Issue of Korean Human Rights During and After the Pacific War." Seoul: Society for the Survivors and the Bereaved Families of the Pacific War (T'aep'yŏngyang chŏnjaeng hŭisaengja yujokhoe), March 1993.

———. "Kŭllo chŏngsindae hyŏnhwang" [Present conditions of the Women's Volunteer Corps]. Unpublished records. Seoul: Society for the Bereaved Families of the Pacific War (T'aep'yŏngyang chŏnjaeng hŭisaengja yujokhoe), 1999.

*Tonga ilbo* [East Asia Daily].

U, Sunok "Yŏgong ilgi" [Diary of a factory girl]. *Pyŏlgŏn'gon* (March 1930). 72–73.

Underwood, Horace H. *Modern Education in Korea* (New York: International Press, 1926).

Underwood, Lillias H. *Fifteen Years Among the Topknots*. New York: American Tract Society, 1904.

———. "Woman's Work in Korea." *The Korean Repository* 3:1 (January 1896): 62–65.

Yi, Chaeyun. Taped Interview conducted by Kang Isu, October 4, 1991.

Yi, Chungnye. Taped Interview, June 3, 1999.

Yi, Kit'aek. *Chŏnnam tongmaeng kangnyŏng* [Platform of the Chŏllanamdo League]. *Shisō geppō* [Review of Thought] 4:1 (April 1934): 36–39.

Yi, Kyŏngsŏn. "Chosŏn yŏsŏng ege hosoham" [A petition for the women of Chosŏn]. *Kaebyŏk* 22:1 (January 1930): 90–96.

Yi, Sŏnghwan. "Puin munje kwanhwa: Puin kwa chigŏp chŏnsŏn" [Problems and solutions to the woman question: Women and the battle of work]. *Sinyŏsŏng* 10:3 (March 1932): 13–19.

Yi, Taeŭi. "Yŏjaŭi kyŏngjejŏk tongnip" [Women's economic liberation]. *Ch'ŏngnyŏn* (January 1922).

*Yŏsŏng* [Woman].

Yu, Kwangnyŏl. "Kyŏlchŏn kungnae t'aese ŭi kanghwa" [The strengthening of wartime national conditions]. *Chogwang* (November 1943): 29–30.

Yuk, Chisu. "Seishi rōdō ni tsuite." *Chōsen sōtokufu chōsa geppō* [Monthly report of the Govenment-General of Korea] 6:7 (July 1938).
Yun, Chohŭi. "Nongch'on kwa puŏp" [Secondary business and the countryside]. *Sinmin* (June 1929): 93–94.


SECONDARY SOURCES

Ackerman, Phillip. "Determinants of Individual Differences During Skill Acquisition: Cognitive Abilities and Information Processing." *Journal of Experimental Psychology: General* 117 (1988): 288–318.
Adams, Hazard, and Searle, Leroy, eds., *Critical Theory Since 1965*. Tallahassee: Florida State University Press, 1986.
Akamatsu, Kaname. "A Historical Pattern of Economic Growth in Developing Countries." *The Developing Economies* 1 (March–August 1962): 3–25.
Alcoff, Linda. "The Problem of Speaking for Others." *Cultural Critique* no. 20 (1991–1992); 5–32.
Amsden, Alice H. *Asia's Next Giant: South Korea and Late Industrialization*. Oxford: Oxford University Press, 1989.
An, P'yŏngjik. "Nihon chissū ni okeru Chōsenjin rōdōsha kaikyū no seicho ni kansuru kenkyū" [Korean Workers and the Japanese Nitrogen Fertilizer Company]. *Chōsenshi kenkyūkai ronbunshū* 25 (1988): 157–92.
———, and Cho, Kijun, eds. *Han'guk nodong munje ŭi kujo* [The structure of the Korean labor movement]. Seoul: Kwangminsa, 1978.
———, and Nakamura, Tetsu, eds. *Kŭndae Chosŏn kongŏphwa ŭi yŏn'gu* [A study of industrialization in modern Korea]. Seoul: Ilchogak, 1993.
———, Yi, Taegŭn, Nakamura, Tetsu, and Kajimura, Hideki, eds. *Kŭndae Chosŏn ŭi kyŏngje kujo* [The economic structure of modern Korea]. Seoul: Pibong ch'ulp'ansa, 1989.
An, P'yŏn'gu, and To, Chinsun, eds. *Pukhanŭi Han'guksa insik* [Korean historiography in North Korea]. Vols. 1 and 2. Seoul: Hangil, 1990.
Armstrong, Charles K. *The North Korean Revolution: 1945–1950*. Ithaca, NY: Cornell University Press, 2003.
Bakhtin, Mikhail M. *The Dialogic Imagination: Four Essays*. Caryl Emerson and Michael Holquist, trans. Austin: University of Texas Press, 1981.
———. *Rabelais and His World*. Helene Isowolsky, trans. Bloomington: Indiana University Press, 1988.
Barbalet, Jack, ed. *Emotions and Sociology*. London: Blackwell, 2002.
Becker, Gary S. "The Economic Way of Looking at Behavior: The Nobel Lecture." *Essays in Public Policy* 69 (Stanford, CA: Hoover Institute, 1996).
———. *A Treatise on the Family*. 1981. Reprint, Cambridge, MA: Harvard University Press, 1991.

Berger, Peter L. *The Capitalist Revolution: Fifty Propositions about Prosperity, Equality, Life and Liberty*. New York: Harper Collins, 1986.

Berlanstein, Lenard L., ed. *Rethinking Labor History*. Urbana: University of Illinois Press, 1993.

Bernstein, Gail Lee, ed. *Recreating Japanese Women, 1600–1945*. Berkeley: University of California Press, 1991.

Bhabha, Homi K. *The Location of Culture*. London: Routledge, 1994.

———. *Nation and Narration*. London: Routledge, 1990.

Bourdieu, Pierre. *The Logic of Practice*. Richard Nice, trans. Cambridge: Polity Press, 1990.

———. *Outline of a Theory of Practice*. Richard Nice, trans. Cambridge: Cambridge University Press, 1977.

Burchell, Graham, Gordon, Colin, and Miller, Peter, eds. *The Foucault Effect*. Chicago: University of Chicago Press, 1991.

Butler, Judith. *Gender Trouble: Feminism and the Subversion of Identity*. New York: Routledge, 1999.

———. *The Psychic Life of Power: Theories in Subjection*. Stanford, CA: Stanford University Press, 1997.

———, Leclau, Ernesto, and Žižek, Slavoj. *Contingency, Hegemony and Universality: Contemporary Dialogues on the Left*. London: Verso, 2000.

———, and Scott, Joan, eds. *Feminists Theorize the Political*. London: Routledge, 1992.

Canning, Kathleen. *Languages of Labor and Gender: Female Factory Work in Germany, 1850–1914*. Ithaca, NY: Cornell University Press, 1996.

Chakrabarty, Dipesh. *Rethinking Working-Class History: Bengal, 1890–1940*. Princeton, NJ: Princeton University Press, 1989.

Chandra, Vipan. *Imperialism, Resistance, and Reform in Late Nineteenth-Century Korea*. Berkeley: University of California Press, 1988.

———. "Sentiment and Ideology in the Nationalism of the Independence Club, 1896–1898." *Korean Studies* 10 (1986).

Chang, Yunshik. "Population in Early Modernization: Korea." Unpublished Ph.D. dissertation, Princeton University, 1967.

Chatterjee, Partha. *The Nation and Its Fragments: Colonial and Postcolonial Histories*. Princeton, NJ: Princeton University Press, 1993.

———, and Jeganathan, Pradeep, eds. *Subaltern Studies XI: Community, Gender and Violence*. New York: Columbia University Press, 2000.

Chen, Edward I-te. "Japanese Colonialism: An Overview." In Harry Wary and Hilary Conroy, eds., *Japan Examined: Perspectives on Modern Japanese History*. Honolulu: University of Hawaii Press, 1983: 201–7.

Cho, Kyung-won. "Critical Examination of the Confucian View of Women's Education." In Jung Chang-young (Chŏng Chang-yŏng), ed., *Korean Social Sciences Journal* 151 (1990).

Cho, Yŏnggu, ed. *Kyŏngsŏng pangjik osimnyŏnsa* [Fifty-year history of the Kyŏng-sŏng Spinning and Weaving Company]. Seoul: Kyŏngsŏng pangjik chusik hoesa, 1969.

Ch'oe, Sŏkno, ed. *Sajin ŭro pon Chosŏn sidae* [Photographs of the Chosŏn era]. Vols. 1 and 2. Seoul: Sŏmundang, 1998.

Ch'oe, Sukkyŏng, Yi, Paeyŏng, Sin, Yŏngsuk, and An, Yŏnsŏn. "Han'guk yŏsŏngsa chŏngnip ŭl wihan yŏsŏng inmul yuhyŏng yŏn'gu: Samil undong ihu put'ŏ hae-bang kkaji" [A study of characteristics of notable women for the strengthening of Korean women's history: From the March First movement to liberation]. *Yŏsŏnghak nonjip*. Vol. 10. Seoul: Ewha Woman's University and Han'guk yŏsŏng yŏn'gusoe, 1993: 11–129.

Ch'oe, Yŏnghŭi, Kim, Sŏngsik, Kim, Yunhwan, and Chŏng, Hosŏp, eds. *Ilcheha ŭi minjok undongsa* [The nationalist movement under Japanese rule]. Seoul: Hyŏnŭmsa, 1982.

Choi, Jang Jip. *Labor and the Authoritarian State: Labor Unions in South Korean Man-ufacturing Industries, 1961–1980.* Seoul: Korea University Press, 1989.

Choi, Sook-kyung. "Formation of the Women's Movement in Korea: From the Enlightenment Period to 1910." *Korea Journal* 25:1 (January 1985).

Chŏn, Sŏktam, and Ch'oe, Unkyu. *Chosŏn kŭndae sahoe kyŏngjesa* [The social and economic history of modern Korea]. 1959. Reprint, Seoul: Yisŏng kwa hyŏn-silsa, 1989.

Chŏn, Sŏktam, Kim, Hanju, and Yi, Kisu. *Ilcheha ŭi Chosŏn sahoe kyŏngjesa* [The so-cial and economic history of Korea under Japanese rule]. Seoul: Hyŏptong mun'go chogŭm yŏnp'an, 1947.

Chŏn, T'aeil. *Nae chugŭm ŭl hŏttoei malla* [Don't waste my life: A collection of es-says]. Seoul: Tolbegae, 1988.

Chŏn, Wuyong. "1930 nyŏndae Chosŏn kongŏphwa wa chungso kongŏp" [Indus-trialization of Chosŏn in the 1930s and small- to medium-sized enterprises]. *Han'guk saron* 25:8 (August 1990): 463–534.

Chŏng, Chinsŏng. "Singminji chabon chuŭihwa kwajŏng esŏ ŭi yŏsŏng nodong ŭi p'yŏnmo" [The transfiguration of female labor during the capitalization process in colonial Korea]. *Han'guk yŏsŏnghak* 4 (1988): 49–100.

Chŏng, Hyegyŏng. "Ilcheha chae-Il Han'gukin minjok undong ŭi yŏn'gu: Taep'an chibang ŭl chungsim ŭro" [A study of the nationalist movements of Koreans in Japan under Japanese rule: A focus on the Ōsaka region]. Unpublished Ph.D. dissertation, Han'guk chŏngsin munhwa yŏn'guwŏn [Academy of Korean Stud-ies], 1998.

Chŏng, Kŭnsik. "Ilcheha Chongyŏn pangjŏk ŭi chamsaŏp chibae" [Labor control of the Chŏngyŏn (Kangegafuchi) silk-reeling factory under Japanese colonial rule]. In *Han'guk kŭndae nongch'on sahoe wa Ilbon cheguk chuŭi* [Rural society and Japanese imperialism in modern Korea], Han'guk sahoesa yŏn'guhoe non-munjip. Vol. 2. Seoul: Munhak kwa chisŏngsa, 1986: 147–94.

Chou, Fang-Lan. "Bible Women and the Development of Education in the Korean Church." In Mark R. Mullins and Richard Fox Young, eds., *Perspectives on Christianity in Korea and Japan: The Gospel and Culture in East Asia.* Lewiston, NY: Edwin Miller Press, 1995.

Chu, Ikchong. "Ilcheha Pyongyang ŭi meriasŭ kongŏp e kwanhan yŏn'gu" [A study of the knitting industry in colonial Pyongyang]. Unpublished Ph.D. dissertation, Seoul National University, 1994.

Chung, Yosup. *Han'guk yŏsŏng undong sa* [History of the Korean women's movement]. Seoul: Ilchŏgak, 1971.

Conroy, Hilary. *The Japanese Seizure of Korea, 1868–1910: A Study of Realism and Idealism in International Relations.* Philadelphia: University of Pennsylvania Press, 1960.

Cook, Haruko Taya, and Cook, Theodore, eds. *Japan at War: An Oral History.* New York: New Press, 1992.

Cott, Nancy F. *The Grounding of Modern Feminism.* New Haven, CT: Yale University Press, 1987.

———, and Pleck, Elizabeth H., eds. *A Heritage of Her Own.* New York: Simon & Schuster, 1979.

de Certeau, Michel. *The Practice of Everyday Life.* Berkeley: University of California Press, 1984.

Deleuze, Gilles. *The Fold: Leibniz and the Baroque.* Tom Conley, trans. Minneapolis: University of Minnesota Press, 1992.

———, and Guattari, Félix. *Anti-Oedipus: Capitalism and Schizophrenia.* Minneapolis: University of Minnesota Press, 1983.

Derrida, Jacques. *Politics of Friendship.* George Collins, trans. London: Verso, 1997.

———."Women in the Beehive: A Seminar with Jacques Derrida." Transcript from the Pembroke Center for Teaching and Research Seminar with Derrida. *Subjects/Objects* (Spring 1984).

Deuchler, Martina. *Confucian Gentlemen and Barbarian Envoys: The Opening of Korea, 1875–1885.* Seattle: University of Washington Press for the Royal Asiatic Society, Korea Branch, 1977.

———. *The Confucian Transformation of Korea: A Study of Society and Ideology.* Cambridge, MA: Harvard University Press, 1992.

Donzelot, Jacques. *The Policing of Families.* Robert Hurley, trans. Baltimore, MD: Johns Hopkins Press, 1979.

Dower, John. *Empire and Aftermath: Yoshida Shigeru and the Japanese Experience, 1878–1954.* Cambridge, MA: Harvard University Press, 1979.

Dublin, Thomas. *Farm to Factory: Women's Letters, 1830–1860.* New York: Columbia University Press, 1981.

———. *Women at Work: The Transformation of Work and Community in Lowell, Massachusetts, 1826–1860.* New York: Columbia University Press, 1981.

Dudden, Alexis. *Japan's Colonization of Korea: Discourse and Power.* Honolulu: University of Hawaii Press, 2005.

Duncombe, Stephen, ed. *Cultural Resistance Reader.* New York: Verso, 2002.

Duus, Peter. *The Abacus and the Sword: The Japanese Penetration of Korea, 1895–1910.* Berkeley: University of California Press, 1995.

———, Myers, Ramon H., and Peattie, Mark R., eds. *The Japanese Wartime Empire, 1931–1945.* Princeton, NJ: Princeton University Press, 1996.

Eckert, Carter J. *Offspring of Empire: The Koch'ang Kim's and the Colonial Origins of Korean Capitalism, 1876–1945.* Seattle: University of Washington Press, 1991.

———. "The South Korean Bourgeoisie: A Class in Search of Hegemony." *Journal of Korean Studies* 7 (1990–91): 115–48.

Engels, Friedrich. *The Origin of the Family, Private Property and the State.* New York: Pathfinder, 1972.

Fanon, Frantz. *Toward the African Revolution.* Harmondsworth, UK: Pelican, 1967.

———. *The Wretched of the Earth.* New York: Grove, 1965.

Firestone, Shulamith. *The Dialectic of Sex: The Case for Feminist Revolution.* New York: Morrow, 1970.

Fortunati, Leopoldina. *The Arcane of Reproduction: Housework, Prostitution, Labor and Capital.* Hilary Creek, trans. New York: Autonomedia, 1995.

Foucault, Michel. *The Archaeology of Knowledge and the Discourse on Language.* Smith, A. M. Sheridan, trans. New York: Pantheon, 1972.

———. *Discipline and Punish: The Birth of the Prison.* New York: Vintage Books, 1979.

———. *The History of Sexuality.* Vols. 1–3. New York: Vintage Books, 1978.

———. *Language, Counter-Memory, Practice: Selected Essays and Interviews.* Donald Bouchard and Sherry Simmons, trans. Ithaca, NY: Cornell University Press, 1977.

———. *Power/Knowledge: Selected Interviews and Other Writings, 1972–1977.* New York: Vintage Books, 1980.

Fox, Richard J., and Starn, Orin, eds. *Between Resistance and Revolution.* New Brunswick, NJ: Rutgers University Press, 1997.

Fujikoshi kōzai kōgyō kabushiki kaisha. *Fujikoshi gojū nenshi* [Fifty-year history of Fujikoshi]. Toyama: Fujikoshi kōzai kōgyō kabushiki kaisha, 1978.

———. *Fujikoshi nijū nenshi* [Twenty-year history of Fujikoshi]. Toyama: Fujikoshi kōzai kōgyō kabushiki kaisha, 1953.

Gerschenkron, Alexander. *Economic Backwardness in Historical Perspective: A Book of Essays.* Cambridge, MA: Belknap Press, 1966.

Giddens, Anthony. *The Consequences of Modernity.* Stanford, CA: Stanford University Press, 1990.

———. *The Constitution of Society: Outline of the Theory of Structuration.* Berkeley: University of California Press, 1984.

———. *Modernity and Self-Identity: Self and Society in the Late Modern Age.* Stanford, CA: Stanford University Press, 1991: 7.

Ginzberg, Carlo. *The Cheese and the Worms: The Cosmos of a Sixteenth-Century Miller.* John and Anne Tedeschi, trans. Baltimore, MD: Johns Hopkins University Press, 1976.

Glenn, Evelyn, Chang, Grace, and Forcey, Linda, eds. *Mothering: Ideology, Experience, and Agency.* London: Routledge, 1994.

Goffman, Erving. *Presentation of Self in Everyday Life.* New York: Doubleday, 1959.

Gordon, Andrew. *The Evolution of Labor Relations in Japan: Heavy Industry, 1853–1955.* Cambridge, MA: Harvard University Press, 1988.

Gouldner, Alvin. *Patterns of Industrial Bureaucracy: A Case Study of Modern Factory Administration.* New York: Macmillan, 1954.

Grajdanzev, Andrew J. *Modern Korea.* 1944. Reprint, New York: Octagon Books, 1978.

Gramsci, Antonio. *Prison Notebooks.* Vols. 1 and 2. New York: Columbia University Press, 1991.

Gruber, Helmut, and Graves, Pamela, eds. *Women and Socialism, Socialism and Women.* Oxford, UK: Berghahn Books, 1998.

Hall, Catherine, and Rose, Sonia, eds. *At Home with the Empire: Metropolitan Culture and the Imperial World.* Cambridge: Cambridge University Press, 2006.

Han, Hŭng-gu, Yi, Chae-hwa, eds. *Chosŏn minjok haebang undongsa charyo ch'ongsŏ.* [The collected records of the Korean movement for national liberation], Seoul: Kyŏngwŏn munhwasa, 1988.

Han'guk chŏngsindae munje taech'aek hyŏpŭihoe [Committee for the Resolution of the Korean Women's Volunteer Corps Problem] and Chŏngsindae yŏnguhoe [Committee for the Research of the Korean Women's Volunteer Corps], eds. *Chungguk ŭro kkŭllyŏgan Chosŏnin kunwianbudŭl* [The forced migration of Korean comfort women to China]. Seoul: Hanŭl, 1995.

————, eds. *Kangjero kkŭllyŏgan Chosŏnin kunwianbudŭl* [The forced migration of Korean comfort women]. Seoul: Hanŭl, 1993.

Han'guk nodong chohap ch'ongyŏnmaeng [Amalgamated Labor Unions of Korea], ed. *Han'guk nodong chohap undongsa* [History of Korean labor unions]. Seoul: Han'guk nodong chohap ch'ongyŏnmaeng, 1979.

————. *Han'guk nodong chohap undongsa* [History of Korean labor unions]. Seoul: Han'guk nodong chohap ch'ongyŏnmaeng, 1991.

————. *Yŏsŏng nodong kwa p'yŏngdŭng* [Equality and female labor]. Seoul: Han'guk nodong chohap ch'ongyŏnmaeng, 1991.

Han'guk sahoesa yŏn'guhoe [Research Institute for Korean Social History], ed. *Han'guk sahoe ŭi yŏsŏng kwa kajok* [Women and family in Korean society], Han'guk sahoesa yŏn'guhoe nonmunjip. Vol. 20. Seoul: Munhak kwa chisŏngsa, 1990.

————. *Ilcheha ŭi sahoe undong* [Social movements in Colonial Korea]. Han'guk sahoesa yŏn'guhoe nonmunjip. Vol. 9. Seoul: Munhak kwa chisŏngsa, 1987.

————, ed. *Ilcheha ŭi sahoe undong kwa nongch'on sahoe* [The social movement of the colonial era and rural society], Han'guk sahoesa yŏn'guhoe nonmunjip. Vol. 25. Seoul: Munhak kwa chisŏngsa, 1990.

Haraway, Donna. "Situated Knowledges: The Science Question in Feminism and the Privilege of Partial Perspectives." *Feminist Studies* 14:3 (Fall 1988): 575–99.

Hareven, Tamara. *Aging and Generational Relations over the Life Course: A Historical and Cross-Cultural Perspective*. New York: W. de Gruyter, 1996.

————. *Families, History and Social Change: Life-Course and Cross-Cultural Perspectives*. Boulder, CO: Westview, 2000.

————. *Family Time and Industrial Time: The Relationship between the Family and Work in a New England Industrial Community*. Cambridge: Cambridge University Press, 1978.

Havens, Thomas R. H. "Women and War in Japan, 1937–1945." *American Historical Review* 80:4 (October 1975): 913–34.

Hershatter, Gail. *The Workers of Tianjin, 1900–1949*. Stanford, CA: Stanford University Press, 1986.

Hŏ, Changman. "1920 nyŏndae minjok kaeryang chuŭi ŭi kaekŭpchŏk kich'o haemyŏng esŏtoenŭn myŏt kaji munje" [Fundamental problems of the progressive nationalism of the 1920s]. *Ryŏksa kwahak* (March 1966): 37–43.

Hŏ, Suyŏl. "Ilcheha Han'guk ŭi ittŏsŏ singmingjijŏk sŏnggyŏk e kwanhan ilyŏngu" [The characteristics of colonial industrialization in Korea under Japanese rule]. Unpublished Ph.D. dissertation, Seoul National University, 1983.

Honey, Maureen. *Creating Rosie the Riveter: Class, Gender and Propaganda During World War II*. Amherst: Massachusetts University Press, 1984.

Honig, Emily. *Sisters and Strangers: Women in the Shanghai Cotton Mills, 1919–1949*. Stanford, CA: Stanford University Press, 1986.

Howard, Judith A., and Callero, Peter, eds. *The Self-Society Dynamic: Cognition, Emotion and Action*. Cambridge: Cambridge University Press, 1991.

Howard, Keith, ed. *True Stories of the Korean Comfort Women*. London: Cassell, 1995.

Hughes, Everett. *The Sociological Eye: Selected Papers*. Vol. 1. Chicago: Aldine-Atherton, 1971.

Hunter, Janet, ed. *Japanese Women Working*. London: Routledge, 1993.

Hwang, Kyung Moon. *Beyond Birth: Social Status in the Emergence of Modern Korea*. Cambridge, MA: Harvard University Press, 2004.

Hwang, Sŏkyŏng. "Kuro kongdan ŭi nodong silt'ae" [The realities of labor in the Kuro industrial area]. *Wŏlgan chungang* (December 1973): 117–27.

Im, Chongguk. *Ilche ch'imnyak kwa ch'in-Ilp'a* [Japanese aggression and collaboration]. Seoul: Ch'ŏngsa ch'ulp'ansa, 1982.

Isogaya, Seiji. *Uri ch'ŏngch'un ŭi Chosŏn: Ilcheha nodong undogn ŭi kirok* [Springtime of Korea: The labor movement under Japanese control]. Kim Kyeil, trans. Seoul: Sakyeŏl, 1988.

Jameson, Frederic. "Postmodernism and Consumer Society." In E. Ann Kaplan, ed., *Postmodernism and Its Discontents: Theories, Practices.* London: Verso, 1988: 13–29

———. *Postmodernism or, The Cultural Logic of Late Capitalism.* Durham, NC: Duke University Press, 1991.

Janelli, Roger, and Dawnhee Yim. *Ancestor Worship and Korean Society.* Stanford, CA: Stanford University Press, 1982.

Janeway, Elizabeth. *Powers of the Weak.* New York: Morrow Quill, 1981.

Janssens, Angelique. *Family and Social Change: The Household as a Process in an Industrializing Community.* Cambridge: Cambridge University Press, 1993.

Jones, Gareth Stedman. *Languages of Class: Studies in English Working Class History, 1832–1982.* Cambridge: Cambridge University Press, 1983.

Joyce, Patrick. *The Rule of Freedom: Liberalism and the Modern City.* London: Verso, 2003.

———, ed. *The Historical Meanings of Work.* Cambridge: Cambridge University Press, 1987.

Kanegafuchi bōseki kabushiki kaisha. *Kanebo hyakunenshi* [One hundred-year history of the Kanegafuchi Spinning and Weaving Company]. Osaka: Kanegafuchi bōseki kabushiki kaisha, 1988.

Kang, Isu. "1920–1960 nyŏn Han'guk yŏsŏng nodong sijang kujo ŭi sajŏk p'yŏn-hwa: Koyong kwa ımgŭm kyŏkch'a p'yŏnhwa rŭl chungsim uro" [Historical changes within the structure of the female labor market in Korea, 1920–1960: Changes in employment and wage differences]. *Yŏsŏng kwa sahoe,* no. 4 (1993): 166–209.

———. "1930 nyŏndae myŏnbang taegiŏp yŏsŏng nodongja ŭi sangt'ae ŭi taehan yŏn'gu" [The conditions of female workers in the cotton textile industry of the 1930s]. Unpublished Ph.D. dissertation, Ewha Woman's University, 1992.

Kang, Kyŏngae. *In'gan munje* [Problems of humanity]. Tonga ilbo serial publications, 1934. Reprint, Seoul: Ch'angjak kwa pip'yŏngsa, 1978.

———, Paek, Sinae, and Kim, Myŏngsun. *Haebang chŏn yŏru chakka sŏnjip* [Selected works of pre-Liberation writers]. Seoul: Pŏmjosa, 1987.

Kang, Tongjin. *Ilbon kŭndaesa.* Seoul: Hangilsa, 1986.

Katznelson, Ira. *Working-Class Formation: Nineteenth-Century Patterns in Western Europe and the United States.* Princeton, NJ: Princeton University Press, 1986.

Kautsky, Karl. *The Social Revolution.* London: Twentieth Century Press, 1902.

Kidd, Yasue Aoki. *Women Workers in the Japanese Cotton Mills, 1880–1920.* Ithaca, NY: Cornell University China-Japan Program, 1978.

Kim, Chin'gyun, and Chŏng, Kŭnsik, eds. *Kŭndae chuch'e wa singminji kyuyul kwŏllŏk* [Modernity and colonial disciplinary power]. Seoul: Munhwa kwa haksa, 1997.

Kim, Chongsŏng. *Han'guk ŭi imgŭm mit nodongja ŭi kwanhan ilyŏn'gu: Ilcheha ŭi nodongja sangt'ae rŭl chungsim uro* [A study of work, wages, and conditions of

Korean workers under Japanese rule]. Pusan: Kyŏnsang taehakkyo inmun sahoe kwahak p'yŏn, 1982.

Kim, Chungyŏl. "Chosŏn pangjik kongjang ŭi chaengŭi" [The factory strike of Chŏsen Spinning and Weaving]. *Nodong kongnon* (January–February 1975): 69–76.

———. "P'yongyang komu kongjang p'aŏp" [The strike of the Pyongyang rubber factory]. *Nodong kongnon* (March 1975): 107–16.

Kim, Hŭiil. "Minjok kaeryang chuŭi ŭi keykŭpchŏk kich'o nŭn yesok purŭjyoji yida" [The class origins of nationalist progressivism is bourgeoise]. *Ryŏksa kwahak* (April 1966): 38–46.

Kim, Hwallan (Helen). "Chigŏp chŏnsŏn kwa Chosŏn yŏsŏng" [Professional progress and the Chosŏn woman]. *Sintonga* 11 (September 1932): 2–9.

———. *Rural Education for the Regeneration of Korea*. Philadelphia: Dunlap Printing Co., 1931.

Kim, Ilmyŏn. *Tennō no guntai Chōsenjin ianfu* [Emperor's forces and Korean comfort women]. Tokyo: Sanchi shobo, 1976.

Kim, Indŏk. *Singminji sidae chaeil Chosŏnin undong yŏn'gu* [A study of the movements of Koreans in Japan during the colonial era]. Seoul: Kukhak charyowŏn, 1996.

Kim, Janice C. H. "The Pacific War and Working Women in Colonial Korea." *Signs* 33:1 (Fall 2007): 81–104.

———. "Processes of Feminine Power: Shamans in Central Korea." In Keith Howard, ed., *Korean Shamanism: Revivals, Survivals and Change*. Seoul: Seoul Press, Royal Asiatic Society, Korea Branch, 1998: 113–32.

———. "Varieties of Women's Work in Colonial Korea." *The Review of Korean Studies* 10:3 (June 2007): 119–46.

Kim, Kyŏngil. *Ilcheha nodong undongsa* [History of labor movements under Japanese rule]. Seoul: Ch'angjakkwa pip'yŏngsa, 1992.

———. *Yi Chaeyu yon'gu: 1930 nyŏndae Seoul ŭi hyŏngmyŏngjŏk nodong undong* [A study of Yi Chaeyu: The revolutionary labor movement of 1930 Seoul]. Seoul: Ch'angjakkwa pip'yŏngsa, 1993.

———, ed. *Puk-Han hakkye ŭi nonong undong yŏn'gu* [North Korean studies of labor and peasant movements]. Seoul: Ch'angjak kwa pip'yŏngsa, 1989.

Kim, Minyŏng. *Ilche ŭi Chosŏnin nodongnyŏk sut'al yŏn'gu* [A study of the exploitation of the Korean labor movement after Liberation]. Seoul: Ilchogak, 1970.

Kim, Nakchung. *Han'guk nodong undongsa, haebanghu p'yŏn* [History of the Korean labor movement after Liberation]. Seoul: Ilchogak, 1970.

Kim, P'yŏngun. *Myŏnbangjŏk sŏmyu kongŏp ch'ongsŏ* [Documents concerning the cotton-spinning fiber industry]. Vol. 1. Seoul: Ŭlyumunhwasa, 1948.

Kim, Seung-Kyung. *Class Struggle or Family Struggle? The Lives of Women Factory Workers in South Korea*. Cambridge: Cambridge University Press, 1997.

Kim, Suhwan, ed. *Tongil pangjik nodong chohap undongsa* [The history of the amalgamated workers of the Tongil Spinning and Weaving Company]. Seoul: Tolbegae, 1985.

Kim, Yŏngmo. "Ilche sidae chiju ŭi sahoejŏk paegyŏng kwa idong" [The social background and mobility of landlords during the Japanese colonial era]. *Asia yŏn'gu* 14 (1971): 107–25.

———. "Kŭnaehwa ŭi sagak: nodong munje" [The dead angle of modernization: The problem of labor]. *Sedae* (November 1971): 169–243.

Kim, Yŏngsŏp. "Hanmal-Ilcheha ŭi chijuje: Kanghwa Kimssiga ŭi ch'usugi rŭl t'onghaesŏ pon chiju kyŏngyŏng" [The landlord system of the late Chosŏn dynasty and under colonial rule: A case study of the Kihanghwa Kim family]. *Tonga munhwa* 11 (November 1972): 3–86.

Kim, Yung-chung (Kim Yungch'ŏng), ed., trans. *Women of Korea: A History from Ancient Times to 1945*. Seoul: Ewha Woman's University Press, Kwang Myŏng Printing Co., 1976.

Kim, Yunhwan. *Han'guk nodong undongsa, Ilcheha p'yŏn* [History of the Korean labor movement under Japanese rule]. Seoul: Ilchogak, 1970.

———. "Kŭndaejŏk imgŭm nodong ŭi hyŏngsŏng kwajŏng" [The formation of modern wage work]. In An P'yŏngjik and Cho Kijun, eds., *Han'guk nodong munje ŭi kujo*. Seoul: Kwangmingsa, 1980: 59–95.

Kobayashi, Hideo. "1930 nendai zenhanki no Chōsen rōdō undō ni tsuite: Heijō gomu kōjō rōdōsha no zenesuto o chūshin ni shite" [The Korean labor movement of the early 1930s: A focus on the Pyongyang rubber workers' general strike]. *Chōsenshi kenkyūkai ronbunshū* 6 (September 1969): 115–22; also reprinted in Namiki Makoto, et al. *1930 nyŏndae minjok haebang undong* [National liberation movements of the 1930s]. Seoul: Kŏrŭm sinsŏ, 1984: 241–69.

———. *Daitōa kyōeiken no keisei to hōkai* [The rise and fall of the East Asian Co-Prosperity Sphere]. Tokyo: Ochanomizu Shobō, 1975.

———. "The Postwar Economic Legacy of Japan's Wartime Empire." In Peter Duus, Ramon H. Myers, and Mark R. Peattie, eds., *The Japanese Wartime Empire, 1931–1945*. Princeton, NJ: Princeton University Press, 1996: 324–34.

Kondō, Kenichi, ed. *Taiheiyō senka no Chōsen* [Korea and Taiwan during the Pacific War]. Vols. 4 and 5. 1961. Reprint, Seoul: Kukhak charyowŏn, 1984.

———. *Taiheiyō senka no Chōsen oyobi Taiwan* [Korea and Taiwan during the Pacific War]. 1961. Reprint, Seoul: Kukhak charyowŏn, 1984.

———. *Taiheiyō senka shūmatsuki Chōsen no chisei* [Policy in Korea in the last period of the Pacific War]. 1961. Reprint, Seoul: Kukhak charyowŏn, 1984.

Kong, Cheuk. *1950 nyŏndae Han'guk ŭi chabonka yŏn'gu* [History of Korean entrepreneurs in the 1950s]. Seoul: Haksul chongsŏ, 1993.

Koo, Hagen, ed. *State and Society in Contemporary Korea*. Ithaca, NY: Cornell University Press, 1993.

Kwak, Kŏnhong. *Ilche ŭi nodong chŏngch'aek kwa Chosŏn nodongja, 1938–1945* [Labor policy under Japanese rule and Korean workers]. Seoul: Sinsŏwŏn, 2001.

Kwŏn, Chungdong. *Yŏja nodong* [Female Labor]. Seoul: Chungang kyŏngjesa, 1986.

Kwŏn, Minjŏng. "Yŏsŏng nodongjae ŭi chikŏpsŏng chilhwan ŭi kwanhan yŏn'gu" [A study of the occupational diseases of women workers]. *Yŏsŏng yŏn'gu* 7:2 (Summer 1989): 94–107.

Kwŏn, T'aeŏk. *Han'guk kŭndae myŏnŏpsa yŏn'gu* [A study of textile industry in Korea]. Seoul: Ilchogak, 1989.

———. "Kyŏngsŭng chingnyu chusik hoesa ŭi sŏngnip kwa kyŏngyŏng" [The establishment and management of the Kyŏngsŏng Cord Company]. *Han'guk saron* 6:12 (December 1979): 300.

Kwŏn, Ŭisik. "Uri nara esŏ nodong kyeg'ŭp hyŏngsŏng kwanjŏng kwa kŭ sigi" [The formation of the working class in our nation and the era]. *Ryŏksa kwahak* (January 1966): 1–9.

Kwŏn, Yŏnguk. "Nihon teikokushūgika no Chōsen rōdō jijō: 1930 nendai o chūshin ni" [Labor conditions in Chōsen under Japanese imperialism: A focus of the 1930s]. *Rekishigaku kenkyū* 303 (August 1965): 25–39.

Lacan, Jacques. *Ecrits*. Bruce Fink, trans. New York: Norton, 2002.

———. *The Four Fundamental Concepts of Psycho-Analysis*. Jacques-Alain Miller, ed., Alan Sheridan, trans. New York: Norton, 1978.

Law, John. *Organizing Modernity*. Oxford: Blackwell, 1994.

Leclau, Ernesto, and Mouffe, Chantel. *Hegemony and Social Strategy: Towards a Radical Democratic Politics*. London: Verso, 1985.

Lee, Ching Kwan. "Familial Hegemony: Gender and Production Politics on Hong Kong's Electronics Shopfloor." *Gender & Society* 7:4 (December 1993): 529–47.

———, ed. *Working in China: Ethnographies of Labor and Workplace Transformation*. New York: Routledge, 2006.

Lee, Ki-baek (Yi Ki-baek), and Wagner, Edward, trans. *A New History of Korea*. Cambridge, MA: Harvard University Press, 1984.

Lee, Kyu-hwan (Yi Kyuhwan). "A Study of Women's Education under Japanese Occupation." *Journal of the Korean Cultural Research Institute of Ewha University* 3 (1962): 145–46.

Lee, Namhee. *The Making of Minjung: Democracy and the Politics of Representation in South Korea*. Ithaca, NY: Cornell University Press, 2007.

Lee, Tae-Young (Yi Tae-yŏng). "Elevation of Women's Rights." *Korea Journal* 4:2 (February 1964): 4–9.

Lerner, Gerda. *The Creation of Feminist Consciousness: From the Middle Ages to Eighteen-Seventy*. New York: Oxford University Press, 1993.

———. *The Creation of Patriarchy*. New York: Oxford University Press, 1986.

Lim, Sun Hee (Yim Sun-hee). "Women and Education in Korea." *Korea Journal* 25:1 (January 1985).

Marx, Karl. *Capital*. Vols. 1–3. London: Penguin, 1976.

Marx, Karl, and Engels, Friedrich. *The Communist Manifesto*. 1888. Reprint, London: Penguin, 1967: 79–90.

———. *The German Ideology.* Vol. 1. New York: International Publishers, 1970.

McNamara, Dennis. *The Colonial Origins of Korean Enterprise, 1910–1945.* Cambridge: Cambridge University Press, 1990.

Meillassoux, Claude. *Maidens, Meal and Money: Capitalism and the Domestic Community.* New York: Cambridge University Press, 1981.

Millett, Kate. *Sexual Politics.* 1970. Reprint, New York: Simon & Schuster, 1990.

Ministry of Labor, Republic of Korea. *Labor Laws of the Republic of Korea 1997.* Seoul: Munwŏn, 1998.

———. *Labor Laws of the Republic of Korea 2005,* http://www.koilaf.org/admin/publication/file/CLS2005.PDF.

Mitchell, Juliet. *Woman's Estate.* New York: Penguin, 1971.

Moore, Barrington. *Injustice: The Social Bases of Obedience and Revolt.* New York: M. E. Sharpe, 1978.

Murphy, James Bernard. *The Moral Economy of Labor: Aristotelian Themes in Economic Theory.* New Haven, CT: Yale University Press, 1993.

Myers, Ramon H., and Peattie, Mark R. *The Japanese Colonial Empire, 1895–1945.* Princeton, NJ: Princeton University Press, 1984.

Negri, Antonio. *Insurgencies: Constituent Power and the Modern State.* Maurizia Boscagli, trans. Minneapolis: University of Minnesota Press, 1999.

Nelson, Cary, and Grossberg, Larry, eds. *Marxism and the Interpretation of Culture.* Chicago: University of Illinois Press, 1988.

Paek, Ugin. "Singminji sidae kyekŏp kujo ŭi kwanhan yŏn'gu" [A study of the class structure during the colonial era]. In Han'guk sahoesa yŏn'guhoe, ed., *Singminji siodae kyekŭp kujo ŭi kwanhan yŏn'gu* [Social class and social change in Korean society], Han'guk sahoesa yŏn'guhoe nonmunjip. Vol. 8. Seoul: Munhak kwa chisŏngsa, 1987: 121–245.

Pahk, Induk. *September Monkey.* New York: Harper & Brothers, 1954.

Pak, Chaeŭl. "Ilcheha myŏnbang chigŏp ŭi sajŏk chŏn'gae" [The historical development of the cotton textile industry under Japanese rule]. Unpublished Ph.D. dissertation, Kyŏnghee University, 1980.

Pak, Hyŏnch'ae, ed. *Han'guk chabonjuŭi wa imgŭm nodong* [Wage labor and Korean capitalism]. Seoul: Hwada, 1984.

Pak, Inhwan, ed. *Kyŏngsŏng pangjik yuksimnyŏnsa* [Sixty-year history of the Kyŏngsŏng Spinning and Weaving Company]. Seoul: Kyŏngsŏng pangjik chusik hoesa, 1980.

Pak, Kyŏngsik. *Chosŏn munje charyo ch'ongsŏ.* Vol. 1: *Chŏnsi kangje yŏnhaeng—nomu kwalli chŏngch'aek* [Collection of sources on problems in Korea. Vol. 1: The displacement of workers during wartime—policies on labor management]. Vols. 1 and 2. Tokyo: Asia mondai kenkyūsho, 1982.

———. *Ilbon cheguk chuŭi ŭi Chosŏn chibae* [The management of Korea during Japanese imperialism]. Seoul: Ch'ŏnga sinsŏ, 1986.

Pak, Maria. *Kiddokkyo wa Han'guk yŏsŏng sasip nyŏnsa* [Forty-year history of Christianity and Korean women]. In *Han'guk yŏsŏng munhwa nonch'ong*. Seoul: Ewha Woman's University Press, 1958.

Pak, Sŏngch'ŏl, ed. *Kyŏngbang ch'ilsimnyŏnsa* [Seventy-year history of Kyŏngbang]. Seoul: Kyŏngsŏng pangjik chusik hoesa, 1989.

Pak, Ŭnsik. "Ilcheha ŭi yŏsŏng nodong undong: Minjŏk ch'abyŏl ŭi sŏngch'abyŏl kkaji kyŏkŏ" [The female labor movement under Japanese rule: The experience of ethnic and sex discrimination]. *Yŏsŏng*, no. 272 (March 1990): 28–31.

Park, Soon-Won. "Colonial Home Front: World War II and Korean Rural Women." Paper presented at the Fifty-fifth Annual Conference of the Association for Asian Studies, New York, March 29, 2003: 1–21.

——. *Colonial Industrialization and Labor in Korea: The Onoda Cement Factory*. Cambridge, MA: Harvard University Press, 1999.

Park, Yong-ock (Pak Yongok). *Han'guk kŭndae yŏsŏng undongsa yŏn'gu* [A study of the history of modern Korean women]. Seoul: Han'guk chongŭm munhwa yŏn'guwŏn, 1984.

——. *Hanguk yŏsŏng hangil undongsa yŏn'gu* [A study of the Korean women's anti-Japanese movement]. Seoul: Chisik sanŏpsa, 1996.

——. "The Women's Modernization Movement in Korea." In Sandra Mattielli, ed., *Virtues in Conflict: Tradition and the Korean Woman Today*. Seoul: Royal Asiatic Society, Korea Branch, Samhwa Publishing Co., 1977.

Patterson, Wayne. *The Korean Frontier in America*. Honolulu: University of Hawaii Press, 1988.

Popkin, Samuel. *The Rational Peasant*. Berkeley: University of California Press, 1979.

Rancière, Jacques. *The Nights of Labor: The Workers' Dream in Nineteenth-Century France*. John Drury, trans. Philadephia: Temple University Press, 1989.

Redding, Paul. *The Logic of Affect*. Ithaca, NY: Cornell University Press, 1999.

Robinson, Michael E. *Cultural Nationalism in Colonial Korea, 1920–1925*. Seattle: University of Washington Press, 1988.

Rodgers, Daniel T. "In Search of Progressivism." *Reviews in American History* 100:12 (December 1982): 113–32.

Rose, Nikolas. *Governing the Soul: The Shaping of the Private Self*. London: Free Association Books, 1999.

Rostow, Walter W. *The Stages of Economic Growth: A Non-Communist Manifesto*. New York: Cambridge University Press, 1990.

Said, Edward. *Culture and Imperialism*. New York: Vintage Books, 1993.

Scheff, Thomas J. *Microsociology: Discourse, Emotion and Social Structure*. Chicago: University of Chicago Press, 1990.

Schmid, Andre. *Korea Between Empires: 1895–1919*. New York: Columbia University Press, 2002.

Scott, James. "Hegemony and the Peasantry." *Politics and Society* (July 1977): 267–96.

————. *The Moral Economy of the Peasant*. New Haven, CT: Yale University Press, 1976.

————. *Weapons of the Weak: Everyday Forms of Peasant Resistance*. New Haven, CT: Yale University Press, 1985.

Scott, Joan W. *Gender and the Politics of History*. New York: Columbia University Press, 1988.

————. *The Glassworkers of Carmaux: French Craftsmen and Political Action in a Nineteenth-Century City*. Cambridge, MA: Harvard University Press, 1974.

————, and Tilly, Louise A. *Women, Work and Family*. New York: Routledge, 1978.

Seki, Keizō. *The Cotton Industry of Japan*. Tokyo: Japan Society for the Promotion of Science, 1956.

Shin, Gi-Wook. *Peasant Protest and Social Change in Colonial Korea*. Seattle: University of Washington Press, 1996.

————, and Robinson, Michael, eds. *Colonial Modernity in Korea*. Cambridge, MA: Harvard University Press, 1999.

Smith, W. Donald. "Beyond 'The Bridge on the River Kwai': Labor Mobilization in the Greater East Asia Co-Prosperity Sphere." *International Labor and Working Class History* 58 (Fall 2000): 219–38.

————. "Korean Women in Prewar Japanese Coal Mining." http://www .econvention.org/imhc/papers/Smith_e_1021.pdf.

Soh, Chunghee Sarah. "From Imperial Gifts to Sex Slaves, Theorizing Symbolic Representations of the 'Comfort Women.'" *Social Science Japan Journal* 3:1 (2000): 59–76.

————. "Human Rights and the 'Comfort Women.'" *Peace Review* 12:1 (January 2000): 123–29.

Somekawa, Ellen, and Smith, Elizabeth. "Theorizing the Writing of History." *Journal of Social History* 22:1 (Fall 1988): 149–61.

Sŏng, T'aegyŏng. "Samil undong sigi ŭi Han'guk nodongja ŭi hwaltong e taehayŏ" [Concerning the activities of Korean workers in the March First era]. *Yŏksa hakpo* 41 (March 1969): 52–83.

Sorensen, Clark W. *Over the Mountains Are Mountains: Korean Peasant Households and Their Adaptations to Rapid Industrialization*. Seattle: University of Washington Press, 1988.

Spencer, Robert. *Yŏgong: Factory Girl*. Seoul: Royal Asiatic Society, 1988.

Spivak, Gayatri Chakravorty. *A Critique of Postcolonial Reason: Toward a History of a Vanishing Present*. Cambridge, MA: Harvard University Press, 1999.

Suh, Dae-sook. *Documents of Korean Communism, 1918–1948*. Princeton, NJ: Princeton University Press, 1970.

————. *The Korean Communist Movement, 1918–1948*. Princeton, NJ: Princeton University Press, 1967.

Suh, Sang-chul (Sŏ Sangch'ŏl). *Growth and Structural Changes in the Korean Economy, 1910–1940*. Cambridge, MA: Harvard University Press, 1978.

Sun, Chŏmsun. *Yŏdŏl sigan nodong ŭl wihayŏ: Hatae chegwa yŏsŏng nodongja dŭl ŭi t'u-jaeng kirok* [For eight hours of work: The records of female workers in the Hatai confectionary company]. Seoul: P'ulbit, 1984.

Sylla, Richard, and Toniolo, Gianni, eds. *Patterns of European Industrialization: The Nineteenth Century*. London: Routledge, 1991.

Taehan pangjik hyŏphoe (Korean Textile Association). *Panghyŏp ch'angnip sipchunyŏn* [The tenth anniversary of the Textile Association]. Seoul: Taehan pangjik hyŏphoe, 1957.

Thompson, Edward P. "Eighteenth-Century English Society: Class Struggle Without Class?" *Social History* 3:2 (May 1978): 163–64.

———. *The Making of the English Working Class*. New York: Pantheon, 1963.

———. *The Poverty of Theory and Other Essays*. New York: Monthly Review Press, 1978.

Tongil pangjik chusik hoesa. *Tongil pangjik sasa, 1955–1981* [History of the Tongil Spinning and Weaving Company]. Seoul: Tongil pangjik chusik hoesa, 1982.

Tōyō bōseki kabushiki kaisha [East Asia Spinning and Weaving Company]. *Tōyōbō hyaku nenshi* [One hundred-year history of Tōyōbō]. Tokyo: Tōyō bōseki kabushiki kaisha, 1986.

Tsurumi, E. Patricia. *Factory Girls: Women in the Thread Mills of Meiji Japan*. Princeton, NJ: Princeton University Press, 1990.

Turner, Victor. *Dramas, Fields, and Metaphors: Symbolic Action in Human Society*. Ithaca, NY: Cornell University Press, 1974.

———, and Bruner, Edward M., eds. *The Anthropology of Experience*. Urbana: University of Illinois Press, 1986.

Wade, Robert. *Governing the Market: Economic Theory and the Role of Government in East Asian Industrialization*. Princeton, NJ: Princeton University Press, 1990.

Wells, Kenneth M. *New God, New Nation: Protestants and Self-Reconstruction Nationalism in Korea, 1896–1937*. Honolulu: University of Hawaii Press, 1990.

Welter, Barbara. "The Cult of True Womanhood: 1812–1860." *American Quarterly* 28 (1966): 151–74.

Wilk, Richard R., ed. *The Household Economy*. New York: Westview Press, 1989.

Williams, Simon. *Emotion and Social Theory: Corporeal Reflections on the Irrational*. London: Sage, 2001.

Wray, Harry, and Conroy, Hilary, eds. *Japan Examined: Perspectives on Modern Japanese History*. Honolulu: University of Hawaii Press, 1983.

Wright, Gordon. *The Ordeal of Total War, 1939–1945*. New York: Harper & Row, 1968: 244.

Yi, Chŏngok. "Ilcheha kongŏp nodong esŏ ŭi minjok kwa sŏng" [Gender and nationality in industrial labor under Japanese rule]. Unpublished Ph.D. dissertation, Seoul National University, 1990.

Yi, Hyojae. *Han'guk ŭi yŏsŏng undong: Ŏje wa onŭl* [The Korean women's movement: Yesterday and today]. Seoul: Chŏngwusa, 1989.

————. *Yŏsŏng ŭi sahoe ŭisik* [Women's social consciousness]. Seoul: P'yŏngminsa, 1980.

Yi, Kyuhŏn, ed. *Sajin ŭro ponŭn tongnip undong* [Photographs of the independence movement]. Vol. 2. Seoul: Sŏmundang, 1996.

Yi, Paeyong. "Han'guk kŭndae sahoe chŏnhwan kwa honin chedo ŭi p'yŏnhwa" [The transformation of modern Korean society and changes in the institution of marriage]. *Ehwa saga yŏn'gu* 23, 24 (1997): 39–54.

————. "Ilche sigi yŏsŏng undong ŭi yŏn'gusŏng gwa kwaje" [The themes and findings of research on the women's movement during Japanese rule]. *Han'guk saron* no. 26 (1992): 251–75.

Yim, Louise (Im Yŏngsin). *My Forty Year Fight for Korea*. Seoul: International Culture Research Center, Chungang University, 1951.

Yŏ, Sunju. "Ilchemalgi Chosŏnin yŏja kŭllo chŏngsindae ŭi kwanhan sil'tae yŏn'gu" [A study of the realities of the Women's Volunteer Corps in the last years of the colonial era]. Unpublished M.A. thesis, Ewha Woman's University, 1994.

Yoo, Theodore Jun. *The Politics of Gender in Colonial Korea: Education, Labor, and Health, 1910–1945*. Berkeley: University of California Press, 2008.

Young, Louise. *Japan's Total Empire*. Berkeley: University of California Press, 1998.

Yun, Chunha, ed. *Onu yŏsŏng ŭi norae* [Songs of working women]. Seoul: Ingan tosŏ ch'ulp'an, 1983.

Yun, P'yŏngsŏk, Sin, Yongha, and An, P'yŏngjik, eds. *Han'guk kŭndae saron*. Vol. 3. Seoul: Chisik sanŏpsa, 1977.

Zaretsky, Eli. *Capitalism, the Family and Personal Life*. New York: Harper & Row, 1976.

Žižek, Slavoj. *The Sublime Object of Ideology*. London: Verso, 1989.

# Index

*Aeguk puinhoe. See* Patriotic Women's
  Society
agency: of factory women, 165–66; intentions and, 165; motivation and, 164
agriculture: challenges for tenant farmers
  in, 53; colonial Korea, policies on, 38;
  commercialization impacting, 40, 54,
  159; Pacific War labor mobilization in,
  138; summer work in, 45; wages, 57.
  *See also* farms
All Labor Movement, 137
All-People Society, 179*n*11
animal husbandry, 56
Appenzeller, Ella, 178*n*9
Asia-Pacific War. *See* Pacific War
authenticity, 162
*Awakening* (Kim Wŏnju), 10

banks: colonial Korea, formation of, 37;
  industrialization, factories' association
  with, 32
Butler, Judith, 17

Cadastral Land Survey, 37; impact of, 53
"Call to Women Workers," 26
carding process, 82
"catch-up" theory, 32
cellular activism, 104–5
Central Alliance for Young Women,
  182*n*27
Chatterjee, Partha, 19
chemical industry: in colonial Korea, 70;
  leading enterprises in, 199*n*41
child labor: increase of, 195*n*10; Japan's
  influence on, 62; laws for mobilizing,

142–43; light industries taking advantage of, 63; Pacific War's demands increasing, 69–70; wages of, 89–90
children: factory women's goals of social
  mobility fulfilled by, 169; family economy, contributions of, 54
China: colonial Korea, overseas migration to, 43–44; factories, Korea's standards v., 31
Chŏn T'aeil, 155
chŏngsindae. *See* volunteer corps
Chŏsen Spinning and Weaving strike:
  demands of, 108; duration of, 107;
  outcome of, 109
Chosŏn dynasty, 32–33
Chosŏn Women Comrades Society,
  182*n*27
Chosŏn Women's Cooperative Society,
  12
Ch'ungmuro shopping district, 43
class: and culture, 19; defining, 14,
  184*n*38; factory women determined by
  labor v., 116, 126; factory women's
  struggles with, 15
colonial Korea: agricultural policies in,
  38; bank formation in, 37; chemical industry in, 70; childbirth/marriage rates
  in, 55; communist and socialist organizations in, 12–13; cotton textile's rise
  in, 66–67; dyeing sector's rise in,
  68–69; economic policies of the
  Government-General in, 37; educational improvements in, 193*n*56;
  female factory labor conditions in, 45;
  factory women, as viewed in, 3–4;

colonial Korea (*continued*)
factory women's states in, 1–2; factory women's protests in, 13–14; female students in, 47–48; feminism v. women's suffrage in, 16; historical records of factory women in, 20–21; household businesses declining in, 39, 50, 57–58; industrialization's pattern in, 154; Japanese cotton textile investment in, 67–68; Japan's currency's in, 37, 177n2; labor division in, 45; labor identity in, 161–62; labor strike motivations in, 162–63; light industries dominated by factory women in, 129; literary contributions of women in, 9–10; mining industry, growth of female workers in, 71–72; modernization impacting labor in, 50; nationalism in, 10–11; overseas migration to China/Hawaii/Japan/Russia during, 43–44; population discrepancies for, 39, 191n41; population growth in, 192n42; property loss in, 52–53; protests after, 156; rubber industry's growth in, 71; rubber industry's priorities in, 70–71; silk reeling industry's rise in, 64, 66; stages of, 209n1; stages of economic development in, 39; transatlantic ideas of women's suffrage adopted by, 184n42; urbanization during, 1, 39–40; USA's percentage of factory women compared to, 103; writing's of women in, 9–10, 180n18

colonization: Korea, evolution of modernization with, 11; process of, 181n22

comfort women, 210n6; Pacific War and, 130

commercialized farming: agriculture affected by, 40; urbanization fueled by, 54, 159

communism: colonial Korea, organizations of, 12–13; labor strikes influenced by, 183n34

commuting: difficulties of, 90–91

Company Law, 190n35

conflict, 174

Confucianism: patrilineal descent system, 77; in education, 141; in factory housing, 80; recruitment using, 62; structure of, 181n24

Constitution (Korea), 6–7

contract labor system, 31

contracts, 87–88

cotton textiles, 79; carding process for, 82; cleaning procedure for, 81; colonial Korea, Japanese investing in, 67–68; colonial Korea, rise in, 66–67; factory conditions for, 91; Fujikoshi steel company's labor standards compared to mills in, 150; household businesses and spinning, 60; Japan's regulation of dust in, 203n30; knitting processes for, 83–84; labor strikes in, 107–10; Mokp'o and Seoul markets for, 41; scutching and mixing process for, 82–83; South Korea, labor strikes in, 156; specialization in producing, 67; weaving process for, 82–83

Crosby, Christina, 17

defensive plans, 78

Derrida, Jacques, 6

discipline: factories' system for, 76, 78; factory women adapting to, 118; factory women's resistance to, 120

domestic service: family economy, unmarried women's role in, 61–62; labor in factories compared to, 118–19; live-in nature of, 196n25; wage discrepancies in, 62

dormitories. *See* factory housing

dyeing industry: colonial Korea rise of, 68–69; emergence of, 201n19; evolution of, 84

East Asia map, 147

East Asian paternalism, 3

Eastern Learning, 7

economy: colonial Korea, developmental stages of, 39; colonial Korea, Government-General's policies motivated by, 37; of factories requiring night shifts, 97; imperial mobilization movement's influence on, 140–41; Korea, development of, 35–36; of Korea affected by Japan's occupation of Manchuria, 45–46; Korea, isolation of, 32; Pacific War and development of,

129; Pacific War spurring regulation of, 132–33. *See also* family economy

education: advances of women in early, 8; colonial Korea, improvement of, 193*n*56; colonial Korea, percentage of girls enrolled in, 47–48; Confucianism in, 141; in factory housing, 119, 208*n*26; family's wealth determining women's, 194*n*57; fencing practice, 139; gender segregation in, 141; imperial mobilization movement in, 141–43; Japanese pressure on families to participate in, 141; Pacific War, women's rise in, 144; positions in the Women's Labor Volunteer Corps dependent on, 146; weaving class in, 48; white-collar positions gained from expansion of, 37. *See also* school housing

Educational Ordinance of Chōsen, 47

Employment Promotion Policy, 133, 212*n*15

England, 33; factory women, France's policies v., 29–30; modernization's prerequisites in, 32

Enlightenment movement, 7

escape: factories, frequency of, 123, 209*n*34; factory women's reasons for, 122–23; factory women's strategies for, 120–21; families dealing with, 121; punishment for, 121–22, 208*n*31; sexual abuse for, 122

factories: China v. Korea, standards for, 31; colonial Korea, conditions of, 45; conditions of cotton textile, 91; disciplinary system of, 76, 78; domestic service compared to labor in, 118–19; escape frequency in, 123, 209*n*34; as families, 76–77; hosiery production of household businesses v., 69; industrialization, banks' association with, 32; Japanese wage scales in, 88–89, 202*n*26; for knitting, labor strikes, 110–11; labor strikes, collaboration between, 112–13; management methods of, 63; night shift requirements of, 97; prison life compared to, 122–23; production-based wage system employed by, 88; recruitment methods of,

87; in rubber industry owned by Japanese investors, 199*n*42; service industry's paternalism compared to, 160; sexual abuse in, 98; skill dividing labor in, 115; South Korea, privatization of, 157; structural differences by industry for, 79; unmarried women recruited by, 62–63

"factory girl," 26

factory housing: benefits of, 49; Confucian paternalism in, 80; education in, 119, 208*n*26; family promised safety in, 94; Fujikoshi steel company, supervisory relations in, 148–49; gender advantages/disadvantages in, 95; cross-cultural differences of, 29; living conditions in, 94; origins of, 93; present state of, 218*n*29; prison-like treatment in, 94, 96; supervisors' roles in, 97

Factory Law of 1911, 204*n*45, 204*n*46

factory women: adaptability of, 49; advantages v. drawbacks confronting, 173; agency of, 165–66; cellular activism of, 104–5; children fulfilling the goals of, 169; class identity struggles of, 15; colonial Korea compared to USA, percentage of, 103; colonial Korea, conditions for, 1–2; colonial Korea, historical records of, 20–21; colonial Korean views of, 3–4; communities formed by, 78, 96, 119–20; competition amongst, 117–18; discipline, adaptation of, 118; discipline, resistance of, 120; distribution by industry for, 64–65; elite feminists' views on gender compared to, 160; emergence of, 171; endurance of, 99–100; England v. France on characteristics of, 29–30; escape, reasons of, 122–23; escape, strategies of, 120–21; family encouragement for, 75; feminism and, 170; feminist, Marxist, nationalist views on, 2; identity debate of family v. individual for, 18; identity issues facing, 4–5; industrialization impacting labor conditions for, 46; Japan v. USA on policies for, 30; Korean Communist Party's goals for, 12–13; labor divided by physical examinations of, 200*n*15;

factory women (*continued*)
   labor impacting the family positions
      of, 152–53; labor strikes in the 1970s
      of, 216n8; labor strikes, offensive
      strategies of, 114; labor strikes over job
      security for, 109–10; labor strikes, soli-
      darity problems of, 164; labor v. class
      in defining, 116, 126; light industry in
      colonial Korea dominated by, 129;
      menstruation problems for, 95; motiva-
      tions of, 131; nationalism and, 102; as
      "new woman," 24–25; night shift's
      dangers/demands on, 98; Pacific War,
      diversity of labor for, 145; Pacific War,
      increasing demand for, 128–29; patri-
      archy challenged by, 174–75; popula-
      tion growth in Seoul attributed to,
      192n46; pregnancy's impact on, 98;
      protests in colonial Korea of, 13–14;
      recess for, 92; recruitment of, 79; silk
      reeling, long hours of work for, 81, 92;
      silk reeling, protest methods of,
      111–12; skill standards, protests of,
      116–17; South Korea, labor strikes of,
      158; spinning process dominated by,
      82; subalternity and, 167–68; supervi-
      sors basing their treatment on the
      physical appearance of, 96–97; tempo-
      rary nature of urban life for, 118;
      tragic portrait of, 27; unions formed
      by, 105–6; wages of, 89–90; wages im-
      pacting identity of, 153–54, 162; work
      evolving into heavy industry for, 28,
      46–49, 70.
factory worker interviews: profiles for,
   21, 23; table for, 22
family: daughters' roles in, 62; education
   of women determined by wealth of,
   194n57; escape and, 121; factories as,
   76–77; factory housing promising
   safety to, 94; factory women encouraged by, 75; Japanese pressure on educational participation for, 141; labor
   changing positions of factory women
   in, 152–53; labor division in, 51–52;
   silk reeling employment introduced by,
   86–87; urbanization splitting, 54–55;
   wages influencing women's roles in,
   52–53; wage work impacting structure
   of, 77–78

family economy: children's contributions
   to, 54; features of, 51; modernization's
   impact on, 72; unmarried women's domestic service in, 61–62
farms: married women's labor duties on,
   55–57; married women's nursery problems on, 57; tenancy on, 53; women at
   work on, 58. *See also* agriculture; commercialized farming
feminism: colonial Korea, women's suffrage v., 16; factory women and, 170;
   identity/gender debate in, 17; Korea,
   socialism v., 9; nationalism linked to,
   178n5; North Korea, socialism leading
   to rise of, 11–13; North Korea/South
   Korea historiography of, 6; patriarchy
   debate in, 16–17; politics, problems
   facing, 17–18, 178n7; Protestantism's
   role in, 7
feminists: factory women, Marxist's,
   nationalist's, views compared to, 2;
   factory women's perspectives of gender
   compared to elite, 160; goals, stages of,
   185n47; labor historiography revised by,
   19; nationalism in organizations of, 7–9
Fifteen Year War. *See* Pacific War
food processing, 58
France, 33; factory women, England's
   policies v., 29–30
Friends of the Rose of Sharon. *See*
   Kŭnuhoe
Foucault, Michel, 34, 76, 78, 204n41
Fujikoshi steel company: cotton textile
   mills' labor standards compared to,
   150; daily routines at, 149–50; establishment of, 146; lunch system at, 149;
   supervisory relations in, factory housing of, 148–49; tool making in, 148

gender: articulations of skill differing by,
   115–16; education segregated by, 141;
   factory housing, advantages/disadvan-
   tages mediated by, 95; factory women's
   views compared to those of elite feminists on, 160; feminism's debate on
   identity and, 17; industrial capitalism,
   labor divisions according to, 2, 27; inheritance determined by, 51; labor
   strikes, impact of, 125, 164
gender equality law, 156

*German Ideology, The* (Marx), 170
Gerschenkron, Alexander, 32
Giddens, Anthony, 35
Government-General of Korea, 20, 34, 128, 132
governor-general, 34, 37
Great Depression, 165; labor strikes stimulated by, 103, 163

Han Kiyŏng, 22, 120, 150, 167
Hawaii, 43–44
heavy industry: defining, 183*n*32; factory women moving from light industry to, 28, 46–49, 70; labor organization of light industry v., 113; Pacific War, labor in, 146; rubber industry's misassociation with, 187*n*63
hosiery production: factories v. household industries, 69; growth of, 84; labor strikes in, 110–11; wage cuts prompting labor strikes in, 207*n*12
household industries: colonial Korea, decline of, 39, 50, 57–58; cotton textile spinning in, 60; hosiery production of factories v., 69; scale of, 58; silk reeling in, 59; textile production in, 59; women transitioning from, 37. *See also* family economy
housing. *See* factory housing; school housing

*I Am Loving* (Kim Myŏngsun), 10
identity: adaptation impacting, 117; colonial Korea, labor as, 161–62; conflict shaping, 174; factory women, family v. individual debate for, 18; of factory women influenced by wages, 153–54, 162; factory women, problems with, 4–5; factory women's struggles with class and, 15; feminism's debate on gender and, 17; Marx on, 170; politics impacting, 34–35, 169; power and, 5; self-esteem, self-efficacy, authenticity shaping, 162; social roles molding, 161, 172; victimhood as, 168
imperial mobilization movement: economic development influenced by, 140–41; in education, 141–43; launching, 137; stages of, 133; structure of, 212*n*19

Independence club, 7, 179*n*11; mission of, 8
Industrial Association Law of September 1938, 132–33
industrial capitalism: gender divisions of labor in, 2, 27; Korea, rise of, 36; labor strikes threatening, 101; Marx/Smith, views on, 32
industrialization: banks/factories associations in, 32; colonial Korea, pattern of, 154; "factory girl" as symbol for, 26; factory women's labor conditions determined by the characteristics of, 46; Korea, WWI impacting, 38; opportunities of, 99; universal models of, 189*n*17. *See also* Meiji industrialization
industry. *See* heavy industry; light industry
inheritance, 51
intelligence, 116

Japan: child labor, influence of, 62; colonial Korea, currency of, 37, 177*n*2; colonial Korea, overseas migration to, 43–44; cotton textile production in colonial Korea, investment of, 67–68; cotton textiles, dust regulation of, 203*n*30; factories, wage scales for workers from, 88–89, 202*n*26; factory women, USA's policies v., 30; families pressured to participate in public education by, 141; Korean economy impacted by Manchurian occupation by, 45–46; Korean politics, rise of, 33–34; Korea, Russia's competition with, 33; Korea's integration into, 34; Major Industries Control Law impacting, 193*n*54; Pacific War's origins with, 128; rubber manufacturing factories owned by, 199*n*42; Shanghai invaded by, 137. *See also* colonial Korea
Japanese Imperial Rule Association, 212*n*19
Japanese National Registration System, 212*n*19
Joyce, Patrick, 173–74

Kaehwa undong. *See* Enlightenment movement
Kang Churyŏng, 155

Kanghwa Treaty, 32–33
Kang Kyŏngae, 10
Kang Pokchŏm, 22, 93, 119, 122–23, 127–28, 166, 173–74
Kim Chŏngmin, 22, 75–76, 141, 173
Kim Chŏngnam, 22, 148–49, 173
Kim, Esther, 7
Kim Il-Sung, 6
Kim Myŏngsun, 10
Kim Ŭnnye, 22, 76, 90, 172–73
Kim Wŏnju, 10, 62, 180n18
Kim Yŏngsŏn, 22, 62
"kin-scription," 54
"kin-time," 54
"kin-work," 54
knitting processes, 83–84; labor strikes, factories in, 110–11. *See also* hosiery production
Korea: colonization/modernization's influence in, 11; economic development of, 35–36; economic isolation of, 32; factories, China's standards v., 31; feminism v. socialism in, 9; industrial capitalism's rise in, 36; Japan, integration of, 34; Japan/Russia's contest over, 33; Japan's Manchurian occupation impacting economy in, 45–46; Japan's rise in politics of, 33–34; silk breeding tradition in, 80–81; WWI impacting industrialization in, 38. *See also* colonial Korea; North Korea; South Korea
Korean Communist Party: collaborations of, 12; creation of, 11; factory women, goals of, 12–13
Korean Labor Association, 133
Korean League for National Spiritual Mobilization, 136
kŭllo chŏngsindae. *See* Labor Volunteer Corps
*Kŭnu*, 26–27
Kŭnuhoe: proposals of, 180n16; mission of, 9, 180n17
Kunze Silk Reeling strike, 97
Kyŏngsŏng League of Female Rubber Workers, 105

labor: colonial Korea, division of, 45; colonial Korea, identity found in, 161–62; evaluating, 124–25; in factories compared to domestic service,

118–19; factories, skill dividing, 115; factory women defined by class v., 116, 126; factory women's physical conditions dividing, 200n15; factory women's positions in family affected by, 152–53; family, division of, 51–52; farms, married women's, 55–57; feminists revising historiography of, 19; Fujikoshi steel company v. cotton textile mills, standards of, 150; in heavy industry during Pacific War, 146; industrial capitalism, gender divisions of, 27; industrialization impacting factory women's conditions in, 46; meaning of, 124; modernization in colonial Korea influencing, 50; Pacific War, chronology of orders for, 134–36; Pacific War, factory women's diversification in, 145; Pacific War mobilizing agriculture and, 138; Pacific War, stages of mobilization of, 133, 136–37, 140–41; South Korea's locality impacting, 172; spinning process, types of, 92–93; unmarried women's motives in, 73–74; Women's Labor Volunteer Corps expanding opportunities for, 152. *See also* child labor; contract labor system
labor strikes: colonial Korea, motivations for, 162–63; communism influencing, 183n34; in cotton textiles, 107–10; factory women fighting for security in, 109–10; of factory women in 1970's, 216n8; factory women's offensive strategies for, 114; factory women's solidarity problems in, 164; gender influencing, 125, 164; Great Depression stimulating, 103, 163; in hosiery production, 110–11; industrial capitalism threatened by, 101; interfactory collaboration for, 112–13; in knitting factories, 110–11; light industry compared to heavy industry, organizing for, 113; Marxists' theories on, 103; nationalism motivating, 102–3; political/moral expressions of, 168–69; in rubber industry, 106–7; in shoemaking, 201n20; in silk reeling, 111–13; skill/value questioned in, 104; South Korea, factory women and, 158; in South Korea's cotton textile mills, 156; tools and

machinery inadequacies leading to, 202$n$25; wage cuts prompting, 207$n$12. *See also* cellular activism; Chosŏn Spinning and Weaving strike; Kunze Silk Reeling strike; Pyongyang Taedong Hosiery Factory strike; Wŏnsan general strike
Labor Volunteer Corps, 130
League of Korean Workers, 11
Lerner, Gerda, 16
life-plans, 78
light industry: child labor exploited by, 63; colonial Korea, factory women dominating, 129; defining, 183$n$32; factory women moving into heavy industry from, 28, 46–49, 70; labor strike organization of heavy industry v., 113
long-range plans, 78
lumber, 58

Major Industries Control Law, 193$n$54
Manchuria, 45–46
married women: commutes of, 90; farms, labor duties for, 55–57; farms, nursery problems for, 57; morning chores of, 90; service industry, rise of, 60–61; shoemaking dominated by, 86
Marxists: factory women, feminist's, nationalist's, views compared to, 2; international/national perspectives of, 3; labor strikes, theories of, 103
Marx, Karl: on identity, 170; industrial capitalism, view of, 32
Meiji empire, 33, 38
Meiji industrialization, 188$n$13
menstruation problems, 95
migration. *See* overseas migration; urbanization
mining industry, 11, 89; colonial Korea, conditions in, 91; wages, agriculture v., 57; women's employment in, 71–72;
mobilization. *See* imperial mobilization movement
Mobilization of War-Related Industries Law, 132
modernization: England's prerequisites for, 32; family economy, impact of, 72; Korea, colonization concomitant with, 11; labor in colonial Korea shaped by, 50; Pacific war accelerating, 151;

women's employment opportunities rising in, 36–37
Mokp'o, 41
morals, 168–69
mulberry field management, 200$n$11

Na Hyesŏk, 9, 180$n$18
Nakamura Takafusa, 38
National General Mobilization Law, 133, 136
nationalism: in colonial Korea, 10–11; factory women and, 102; feminist organizations coupled with, 7–9; labor strikes motivated by, 102–3
nationalists: factory women, feminist's, Marxist's, views compared to, 2; feminism linked to, 178$n$5
National Mobilization Movement, 137
National Registration System, 137
"new woman," 24–25
night shift: factories requiring, 97; Factory Law of 1911 impacting, 204$n$45, 204$n$46; factory women, dangers/demands of, 98
North Korea: feminism's historiography, South Korea compared to, 6; gender equality law in, 156; socialism and feminism in, 11–13
North Korean Federation of Trade Unions, 156

"oath as subjects of the imperial nation," 140
overseas migration, 43–44

Pacific War: agriculture labor mobilized in, 138; child labor increased due to, 69–70; chronology of labor-related orders for, 134–36; comfort women in, 130; demand for factory women increased in, 128–29; economic development, population growth, urbanization in, 129; economic regulations and, 132–33; education, women's participation rising during, 144; factory women's labor diversification during, 145; heavy industrial labor in, 146; Japan and origins of, 128; modernization accelerated by, 151; Pearl Harbor attack intensifying recruitment in, 137;

Pacific War (*continued*)
professional mobility during, 166; social hierarchy emerging from, 151–52; stages of labor mobilization during, 133, 136–37, 140–41; Women's Labor Volunteer Corps' opportunities during, 144–45
"Paek-Wood Agreement," 157
Pak Hojin, 26
Pak Sun'gŭm, 22, 167
Park, Soon-Won, 101–2
patriarchy: factory women challenging, 174–75; feminism, debate over, 16–17; property and, 17
patriot day, 140
Patriotic Corps of Industrial Workers, 137
Patriotic Labor Association Law, 137
Patriotic Women's Society, 8–9, 180n15
"patriot's marching song," 140
Pearl Harbor attack, 127; Pacific War recruitment intensified after, 137
penalties. *See* wage penalties
perceptual speed, 116
piecers, 93, 200n14
plans. *See* defensive plans; life-plans; long-range plans
Policy on the Regional Regulation of Labor Supply, 87
politics: feminism's problems with, 17–18, 178n7; identity, impact of, 34–35, 169; Japan's rise in Korea and, 33–34; labor strikes, expressions of, 168
Popkin, Samuel, 73
population: colonial Korea, discrepancies on, 39, 191n41; colonial Korea, growth of, 192n42; Pacific War and growth of, 129; Seoul, downtown shaped by growth of, 42; Seoul, factory women causing growth of, 192n46; urbanization impacting, 40, 43
power: circumstance confining, 117; identity and, 5
Practical Learning, 7
Praise and Encouragement Society, 179n11; women's rights promoted in, 8
pregnancy, 98
prison: factories compared to life in, 122–23; factory housing, treatment compared to, 94, 96
privatization, 157

*Problems of Humanity* (Kang Kyŏngae), 10
production-based wage system: factories employing, 88; supervisors determining, 96
propaganda. *See* wartime propaganda
property: colonial Korea, challenges concerning, 52–53; patriarchy and, 17
Protestantism, 15–16; conversions to, 3; feminism aided by, 7
protests: of Chŏn T'aeil, Kang Churyŏng, 155; after colonial Korea, 156; of factory women in colonial Korea, 13–14; in silk reeling, strategies of factory women, 111–12; on skill standards by factory women, 116–17; on wages, 89. *See also* labor strikes; religious activism
Provision for Emergency Student Mobilization, 142
Provision for Emergency War Operations, 142
Provision for the Establishment of Wartime Student Organizations, 141–42
psychomotor ability, 116
Pyongyang Amalgamated Rubber Worker's Union, 106–7
Pyongyang League of Young Women Hosiery Workers, 105
Pyongyang Taedong Hosiery Factory strike, 110

radio, 150
recess, 92
recruitment: Confucian paternalism used in, 62; Employment Promotion Policy for, 212n15; factories, methods for, 87; of factory women, 79; Pacific War, Pearl Harbor attack intensifying, 137; of unmarried women by factories, 62–63
religious activism, 178n9
Republic of Korea, 157
Rose, Nikolas, 34–35
rubber industry, 79; colonial Korea, growth of, 71; colonial Korea, priorities of, 70–71; gender affecting placement in, 85–86; growth rate of, 199n43; heavy industry, mislabeling of, 187n63; Japanese owned factories in, 199n42; labor strikes in, 106–7
runaways. *See* escape

Rural Revitalization Campaign, 56
Russia: colonial Korea, overseas migration to, 43–44; Korea, Japan fighting with, 33

Sang-Chul Suh, 68
school housing, 180*n*13
Scott, James, 73
Scranton, Mary, 178*n*9
scrutching and mixing process, 82–83
self-efficacy, 162
self-esteem, 162
Seoul: Ch'ungmuro shopping district in, 43; cotton textile market in, 41; factory women causing population growth in, 192*n*46; population density changing downtown, 42
service industry: factories' paternalism compared to, 160; married women working in, 60–61
sexual abuse: escape and, 122; in factories, 98. *See also* comfort women
Shanghai, 88, 146; contract labor system in, 31; Japan's invasion of, 137
shoemaking: labor strikes in, 201*n*20; married women dominating, 86; rubber industry, gender division of labor in, 85–86
silk breeding: Korea, tradition of, 80–81; mulberry field management in, 200*n*11
silk reeling, 59, 79; colonial Korea, rise of, 64, 66; factory women's long hours for, 81, 92; family introducing employment in, 86–87; labor strikes in, 111–13; process of, 81; protest methods of factory women in, 111–12
Sirhak. *See* Practical Learning
skill: defining, 114–15; factories dividing labor by, 115; factory women protesting standards of, 116–17; gender differences, proclaiming, 115–16; intelligence/perceptual speed/psychomotor ability comprising, 116; labor strikes questioning, 104; Pacific War, attainability of, 166
Smith, Adam, 32
social hierarchy, 151–52
socialism: colonial Korea, organizations of, 12–13; Korea, feminism v., 9; North Korea, feminism rising from, 11–13

Socialist League of Korean Women, 9
social roles, 161, 172
Sorensen, Clark, 56
South Korea: factory women's labor strikes in, 158; historiography of feminism, North Korea compared to, 6; labor policies affected by locality in, 172; labor strikes in cotton textile mills of, 156; privatization of factories in, 157; USAMGIK taking over businesses in, 157
speed. *See* perceptual speed
spinning process: dust problems in, 91; factory women dominating, 82; labor types in, 92–93; novices in, 84; piecers' responsibilities in, 93
Spivak, Gayatri, 4
steel industry. *See* Fujikoshi steel company
strategies: of factory women for escape, 120–21; labor strikes, factory women's offensive, 114; tactics compared to, 77
strikes. *See* labor strikes
subalternity, 167–68
supervisors: duties of, 204*n*42; factory housing, roles of, 97; factory women's appearance determining treatment by, 96–97; Fujikoshi steel company's factory housing, relations with, 148–49; production-based wage system determined by, 96

tactics, 77
Temporary Fund Control Law, 132
Temporary Regulations for Imports and Exports Law, 132
tenant farming, 53
textiles. *See* cotton textiles; silk reeling
Thompson, E. P., 14
Tonghak. *See* Eastern Learning
Tongnip hyŏphoe. *See* Independence club
tool making, 148
Treaty of Annexation, 33
Turner, Victor, 104

United States Army Military Government in Korea (USAMGIK), 156–57
United States of America (USA): colonial Korea, adopting women's suffrage ideas of, 184*n*42; colonial Korea's

United States of America (*continued*)
percentage of factory women compared to, 103; factory women, Japan's
industrial policies v., 30
unmarried women: factories recruiting,
62–63; family economy, domestic service role of, 61–62; labor, motives of,
73–74; urbanization of, 159; wage distribution among, 73, 79
urbanization: in colonial Korea, 1, 39–40;
commercialized farming fueling, 54,
159; factory women, temporary nature
of, 118; families separated by, 54–55;
Pacific War and, 129; population
growth impacted by, 40, 43; of unmarried women, 159

vegetable cultivation, 56–57
volunteer corps, 130
vulcanization process, 85–86

wage penalties, 87–88
wages: agriculture v. mining industry, 57;
domestic service, discrepancies in, 62;
factories, Japanese exploitation of,
88–89, 202n26; factory women/child
labor, exploitation of, 89–90; factory
women's identities influenced by,
153–54, 162; family roles of women affected by, 52–53; family structure impacted by work for, 77–78; labor strikes
of hosiery workers caused by cuts to,
207n12; protests over, 89; unmarried
women's distribution of, 73, 79. *See also*
production-based wage system
wartime propaganda, 151

weaving process, 82–83; room in, 109
women. *See* comfort women; factory
women; married women; unmarried
women
Women's Foreign Missionary Society, 7
Women's Labor Volunteer Corps: admission standards for, 144; creation of,
143; drafting of, 143–44; education determining position in, 146; information
lacking on, 130–31; labor opportunities
expanded by, 152; Pacific War, opportunities provided by, 144–45; selection
of candidates for, 145
women's rights, 8
Women's Society for Korean Independence, 8–9
women's suffrage: colonial Korea adopting transatlantic ideas for, 184n42;
colonial Korea, feminism and, 16;
Constitution's enactment leading to,
6–7; origins of, 178n8
"Women Workers," 4
Wŏnsan general strike, 101–2
World War I (WWI), 165–66; investment opportunities of, 128; Korea,
industrialization launched by, 38
World War II (WWII), 150

Yi Chaeyun, 22, 98, 121–22
Yi Chungnye, 22, 75, 173–74
Yŏja kŭllo chŏngsindae. *See* Women's
Labor Volunteer Corps
Yu Kwangnyŏl, 144
Yushin Reforms, 216n8

Zaretsky, Eli, 18